Advanced Techniques for Optimal Sizing of Analog Integrated Circuits

IEEE Press
445 Hoes Lane
Piscataway, NJ 08854

Advanced Techniques for Optimal Sizing of Analog Integrated Circuits

Quantum Computing, Machine Learning, and Bio-inspired Optimization

Trang Hoang
Ho Chi Minh City University of Technology (HCMUT), Vietnam

Thinh Quang Do
Memorial University of Newfoundland, Canada

Thang Quoc Nguyen
Memorial University of Newfoundland, Canada

Hoang Trong Nguyen
Memorial University of Newfoundland, Canada

Lihong Zhang
Memorial University of Newfoundland, Canada

Octavia A. Dobre
Memorial University of Newfoundland, Canada

Trung Q. Duong
Memorial University of Newfoundland, Canada

IEEEPress
WILEY

Published by John Wiley & Sons, Inc., Hoboken, New Jersey.

Library of Congress Cataloging-in-Publication Data Applied for:

Hardback ISBN: 9781394296231

Cover Design: Wiley
Cover Image: © Berkah/Getty Images

Set in 9.5/12.5pt STIXTwoText by Straive, Chennai, India

The manufacturer's authorized representative according to the EU General Product Safety Regulation is Wiley-VCH GmbH, Boschstr. 12, 69469 Weinheim, Germany, e-mail: Product_Safety@wiley.com.

C9781394296231_221025

Contents

About the Authors

Trang Hoang is an Associate Professor in Electronics Engineering at Ho Chi Minh City University of Technology (HCMUT), Vietnam National University Ho Chi Minh City (VNU-HCM), Vietnam, and a Dean of the Graduate School, HCMUT, VNU-HCM. His research interests are in the domains of Application-Specific Integrated Circuit (ASIC)/Field-Programmable Gate Array (FPGA) implementation, integrated circuit (IC) architecture, Micro-Electro-Mechanical Systems (MEMS), wireless communications, wireless security, signal processing, optimization for IC design and wireless networks, artificial intelligence (AI), and quantum computing.

Thinh Quang Do is currently working toward his PhD at Memorial University, Canada. His research fields include IC design, machine learning, encryption/decryption, quantum algorithms, and computer arithmetic.

Thang Quoc Nguyen is currently working toward his PhD at Memorial University at Memorial University, Canada. His research interests include analog integrated circuit design and optimization.

Hoang Trong Nguyen is pursuing the Master's degree in Electrical Engineering at Memorial University, Canada. His fields of research interest are analog IC design, optimization, AI, machine learning, and quantum machine learning.

Lihong Zhang is a Full Professor in the Department of Electrical and Computer Engineering, Faculty of Engineering and Applied Science, Memorial University, Canada. His current research interests include very large-scale integration computer-aided design, mixed-signal integrated system/circuit design, microelectromechanical systems design and design automation, wireless sensor networks, microfluidics and biosensors, and microprocessor-based instrumentation for ocean and biomedical applications.

Octavia A. Dobre is a Professor and Tier-1 Canada Research Chair with Memorial University, Canada. Her research interests encompass wireless communication and networking technologies, as well as optical and underwater communications. She has (co-)authored over 500 refereed papers in these areas and obtained 10 Best Paper Awards including the IEEE Heinrich Hertz Award. She serves as the VP Publications of the IEEE Communications Society. She was the inaugural Editor-in-Chief (EiC) of the *IEEE Open Journal of the Communications Society* and the EiC of the *IEEE Communications Letters*. Dr. Dobre is an elected member of the European Academy of Sciences and Arts, a Fellow of the Engineering Institute of Canada, a Fellow of the Canadian Academy of Engineering, and a Fellow of the Royal Society of Canada.

Trung Q. Duong is a Canada Excellence Research Chair (CERC) and a Full Professor at Memorial University, Canada. He is also an Adjunct Professor at Queen's University Belfast, United Kingdom, and a Visiting Professor at Kyung Hee University, South Korea. His current research interests include wireless communications, quantum machine learning, and quantum optimization. He has received two prestigious awards from the Royal Academy of Engineering (RAEng): RAEng Research Chair in 2020 and RAEng Research Fellow in 2015. He is the recipient of the prestigious Newton Prize 2017. He is a fellow of IEEE, EIC, IET, and AAIA. He is currently the Editor-in-Chief of *IEEE Communications Surveys & Tutorials*.

Preface

Integrated circuits (ICs) play a crucial role in modern technology, transforming every aspect of our daily lives. ICs enable compact and high-performance devices such as smartphones, laptops, and home appliances, while underpinning advances in communication technologies including the fourth-/fifth-generation (4G/5G) wireless networks, Wi-Fi, and Bluetooth technologies. They are the driving force behind Internet of Things (IoT) devices for smart homes, medical devices such as magnetic resonance imaging scanners and pacemakers, and autonomous driving technology, to name but a few. In addition, ICs are the foundation for artificial intelligence/cloud computing and data centers, fueling computational advancements. Thanks to their ability to improve efficiency, minimize power consumption, and facilitate technological innovations, ICs remain essential in our modern life.

The IC design process consists of three main stages: logic, analog, and physical designs, among which analog IC design contributes significantly to noise cancellation and reduction in both power consumption and chip area. In analog IC design, proper sizing of transistors and passive components plays a critical role in determining a circuit's performance: their optimal physical dimensions directly affect the circuit's key parameters such as gain, bandwidth, speed, power dissipation, stability, and noise characteristics. Without suitable component sizing, even a well-conceived circuit architecture may fail to meet design specifications, leading to poor performance and hence increased manufacturing costs. Therefore, meticulously sizing in analog IC design is not only a technical necessity but also a crucial factor in achieving high performance, reliability, and efficiency in IC.

Analog sizing procedure involves determining the optimal dimensions of transistors and circuit components to achieve targeted performance for design parameters. This process mirrors the evolution of biological societies, including humans and animal species, where individuals and groups adapt their structures and behaviors to survive and enhance their overall efficiency. While social evolution fine-tunes cooperation, specialization, and resource distribution to enhance group success, analog sizing balances the trade-offs between circuit parameters. Moreover, the sizing procedure in IC design is typically implemented hierarchically, from the smallest building blocks to the top-level design. In the same manner, evolution and natural adaptations often emerge at the smallest scale, with genetic mutations and adaptations at the individual level, which then gradually influence larger social structures over subsequent generations concerning the development of traits and behaviors that shape the dynamics of entire societies. In other words, both processes rely on a bottom-up approach, where local refinements contribute to the overall efficiency and functionality of the whole system.

The study of sizing automation can be further developed for even higher levels of intellectual evolution. Human beings are at the highest level of intelligence, and our brain functions are vaguely and incompletely defined. Machine learning, with neural networks as the core structure, is endeavoring to imitate our networks of neurons to gradually learn how to solve problems efficiently and intuitively, and therefore can be a valuable assistant in the complex sizing design space. Quantum computing, endowed with truly natural behaviors of atoms, is capable of reconstructing any kind of physical system with high precision. Many quantum equations are derived to solve biomedical, financial issues, and other various types of problems. Sizing is not out of this scope and can be formulated to fit quantum computing to overcome high-dimensional data and complex search spaces.

In this book, we present a systematic overview of the challenges of the analog circuit sizing design problems and the principles behind the algorithms suitable for solving it. These algorithms start with the evolution of micro-level biological structures of genes and chromosomes, then continue to ladder up to the social behaviors of the animal population. Those of human populations are also analyzed before we go back to the function of human brains and finish with atom-level again based on the quantum computing fundamentals. Examples of analog circuits are also given to demonstrate the feasibility of those algorithms on the circuit sizing design. Some algorithms can be irrelevant at first, but prove to be useful in specific applications. Throughout theoretical and practical demonstrations, we hope to provide readers with knowledge and technical skills to develop innovations on analog circuit sizing, one major part in the IC design industry.

This book is organized into six chapters. Chapter 1 provides an overview of general IC design optimization, followed by the problem modeling and procedure of analog circuit sizing optimization. Subsequent to introducing these basics, the following chapters illustrate the utilization of various optimization techniques for analog sizing optimization with several case studies for further clarification of the optimization process. In particular, meta-heuristic algorithms, categorized into evolutionary, animal-behavior-inspired, and human-behavior-inspired methods, are discussed in Chapters 2, 3, and 4, respectively. Throughout these chapters, several case studies of widely used analog circuits such as the comparator, bandgap reference, and operational amplifier circuit are presented for further clarification of the optimization process. Subsequently, Chapter 5 delves into the field of machine learning together with its state-of-the-art techniques such as neural network, hyperparameter search, and deep reinforcement learning for optimal sizing of analog circuits, followed by the case study of the analog-to-digital converter. Finally, the book concludes with Chapter 6, regarding quantum machine learning methods which combine the principles of quantum computing and machine learning, for analog sizing optimization.

2nd September 2025

Trang Hoang
Ho Chi Minh City
University of Technology (HCMUT)
Vietnam

Acronyms

6G	sixth generation
ACO	ant colony optimization
ADC	analog-to-digital converter
AFT	Ali Baba and the forty thieves
AHQGA	Auto-adjusting hybrid quantum genetic algorithm
AI	artificial intelligence
BA	bat algorithm
BBO	biogeography-based optimization
BERT	bidirectional encoder representations from transformers
BGR	bandgap reference
CMOS	complementary metal–oxide–semiconductor
CMRR	common-mode rejection ratio
CPU	central processing unit
CS	common source
CS	cuckoo search
DA	drawer algorithm
DAC	digital-to-analog converter
DE	differential evolution
DRL	deep reinforcement learning
EDA	electronic design automation
FA	firefly algorithm
FinFET	fin field effect transistor
FoM	figure of merit
FPA	flower pollination algorithm
GA	genetic algorithm
GPSOA	gravitational particle swarm optimization algorithm
GPT	generative pre-trained transformer
GSA	gravitational search algorithm
HSI	habitat suitability index

HQFGA	hybrid quantum firefly–genetic algorithm
IC	integrated circuit
IoT	Internet of Things
KCL	Kirchhoff's current law
LB	lower bound
ML	machine learning
MOSFET	metal–oxide–semiconductor field effect transistor
NLP	natural language processing
OTA	operational transconductance amplifier
PDP	power-delay product
PM	phase margin
PO	political optimizer
PPO	proximal policy optimization
PSO	particle swarm optimization
PSO-GEO	particle swarm optimization—global exploitation orienter
PSO-LEO	particle swarm optimization—local exploitation orienter
PVT	process, voltage, and temperature
QA	quantum annealing
QGA	quantum genetic algorithm
QML	quantum machine learning
QUBO	quadratic unconstrained binary optimization
RF	radio frequency
RL	reinforcement learning
SaAFT	self-adaptive Alibaba and forty thieves' algorithm
SaDE	self-adaptive differential evolution
SGD	stochastic gradient descent
SIV	suitability index variable
SR	slew rate
UB	upper bound
UGB	unity gain bandwidth
WSO	war strategy optimization

1

Overview and Problem Formulation

Circuit sizing is one of the three main phases of analog circuit design, which contributes enormously to the development of integrated circuit (IC) technology. This chapter discusses the importance of circuit sizing as well as the various challenges that designers might face in handling this task. The general procedure and problem formulation for sizing design, which involves multiple constraints and trade-offs, are also presented.

1.1 Integrated Circuit Design Optimization

In recent decades, integrated circuits (ICs) have played an essential role in modern technology as the fundamental building blocks of electronic devices. Consisting of numerous basic elements called transistors that behave as controlled switches, ICs have been able to form almost any kind of functional device that meets users' needs. Based on their material and operation, transistors can have many different forms and names, among which metal–oxide–semiconductor field effect transistor (MOSFET) is the most commonly used. The advent of complementary metal–oxide semiconductor (CMOS) technology, which creates pairs of n-type and p-type MOSFETs in circuit logic, has offered significant improvement to the integrity of ICs, particularly analog ICs, thanks to the ability to cancel noise, reduce static power consumption, and occupy less space. This has paved the way for the revolution of the IC industry, enabling the development of high-quality computers, smartphones, and countless other gadgets that transform the way we live and work. As a result, CMOS technology is viewed as the dominant technology in the field of microelectronics. Sizing, which indicates the direct optimization of the physical sizes of ICs' transistors, has since been thoroughly studied to satisfy the ever-increasing demand of the electronics market and the exponential growth in the number of transistors fabricated on a chip to enhance

Advanced Techniques for Optimal Sizing of Analog Integrated Circuits: Quantum Computing, Machine Learning, and Bio-inspired Optimization, First Edition.
Trang Hoang, Thinh Quang Do, Thang Quoc Nguyen, Hoang Trong Nguyen, Lihong Zhang, Octavia A. Dobre, and Trung Q. Duong.

its density and performance. In fact, as of 2020, the deep learning processor Wafer Scale Engine 2 by Cerebras has 2.6 trillion MOSFETs on the TSMC 7 nm FinFET process. Moreover, the average number of transistors on a single chip nowadays has reached billions. Dealing with this huge number of transistors extends the search space to such a great extent that it poses serious challenges to the scalability and reliability of the circuit design flow [1].

In addition to the enormous search space, turnaround time should be another factor to consider. The increase in chip density resulting from the decrease in the transistor's dimension leads to a remarkably higher level of complexity in both the semiconductor process and design constraints. The growing demand for high-quality ICs to meet modern requirements, such as high accuracy, low latency, and high speed, further intensifies the intricacy and complexity of the design. Electrical design automation (EDA) has been applied in the IC design procedure, where technical simulation and evaluation tools are implemented to automate the process. Consequently, the turnaround time can be shortened considerably, though proper manipulation and algorithms/techniques for EDA may not be available or usable as the scale of these ICs continues to increase. For these very high-scale level ICs, it should take a long time for conventional EDA to obtain an optimal solution. In the worst case, a solution cannot be found and the designers are required to relax their constraints on efficiency. A flaw within the design procedure, which involves many decisive steps, can cause postponement, delay, or low-quality products. Since the turnaround time of a chip is affected by the efficiency of the EDA [2], the growth in the number of transistors per chip may delay time-to-market with undesirable impacts.

To address the aforementioned issues regarding the enormous search space and time to market, there arises the need for more advanced design techniques as support for the available EDA. Artificial intelligence (AI), with its subset machine learning (ML), has been utilized in a variety of areas. Instead of exploring the design search space in a random or unstructured order as traditional methods, AI/ML solutions are oriented based on previous data and reasonable predictions. In other words, AI/ML algorithms can identify data patterns and make relevant decisions accordingly [2]. Furthermore, thanks to the ability to process vast amounts of data simultaneously and explore multiple solutions in a parallel manner, quantum computing has also attracted the attention of both industry and academia. Therefore, AI, ML, and quantum computing should be promising candidates to handle the obstacles in the development of IC design.

1.2 The Need for Analog IC Design Optimization

We are witnessing how wireless communication, thanks to the development of high-speed wireless links such as Bluetooth Low Energy (BLE), Wi-Fi, ZigBee, and LoRa, has reached a level of ubiquity. This is comparable to the state of electrical power distribution, which facilitates seamless connections among various electronic devices, including laptops, cameras, phones, and domestic appliances [3]. This surge in wireless communication can be attributed to several converging factors. Primarily, the continuous advancement in the quantity and quality of electronic circuits and components has played a pivotal role in this development process. In fact, recent advances in CMOS technology have significantly enhanced the integrity of CMOS ICs, mainly analog ICs, which are responsible for processing and transmitting analog signals, especially those in the radio frequency (RF) spectrum. Analog circuits, which process the noisy continuous signals into digitized ones, are usually built upon complex mathematical models to cope with the noise, process variations, nonlinear effects, and parasitic components. As a result, the scale of analog ICs is relatively small, and their automation is somewhat limited due to a large number of variables and computations. Moreover, maintaining integrity enhancement along with the growing demand for high-quality ICs to meet modern communication requirements, such as ultra-high accuracy, ultra-low latency, and ultra-high speed requirements in sixth-generation (6G) network communication [4], is making the design of analog circuits increasingly intricate and complex.

In digital circuit design, the design automation procedure has matured, as can be seen by the development of hardware description languages (HDLs) such as VHDL, Verilog, and SystemVerilog, and the widespread application of Electronic Design Automation (EDA) tools from Cadence, Synopsys, and Siemens. In contrast, analog IC design automation is still an open research area. This disparity between digital and analog design can be explained by the difficulty in abstracting standardization of the analog circuit procedure due to the continuity of the analog signal. In digital circuit design, the signal is a sequence of two logical levels, 0 and 1, and the specifications are functions of two logical levels only. Based on the circuit behavior described in the form of HDL, the EDA tools synthesize the digital circuit in the form of functional blocks, then realize these functional blocks by using the standard cells provided by the manufacturer. In contrast, the continuity of the analog signal leads to the diversity of the specifications of an analog IC, resulting in a broader design space of device size and topology, which the standard cell library cannot cover. Moreover, continuous analog signal metrics are more complex to reach and at lower accuracy than digital signals, which have already been quantized. Therefore, in digital circuit physical design, mathematical equations can be applied to guide the computer on how to do placement and routing to satisfy some constraints, while in analog IC design, a similar concept is hard to apply. Hence, the achievement in analog IC design automation is still limited. Analog IC design is still heavily manual in a time-consuming and labor-intensive design cycle that negatively impacts IC production businesses, causing potential market share losses due to delayed product launches. In fact, as highlighted in [5], IC design businesses can lose up to 14% of market shares if products are introduced four weeks late.

The process of designing analog ICs follows a structured sequence comprising three main phases: topology design, transistor sizing, and physical design (layout). The first phase involves converting the specified design requirements into a circuit structure. This task strongly depends on the designer's expertise in selecting and connecting suitable components such as MOSFETs, resistors, inductors, and capacitors, into a unified circuit that could reach the required specifications. Subsequent to the topology design is the crucial phase of transistor sizing, where the designer meticulously adjusts the width and length of each transistor in the selected architecture. It is worth mentioning that although initial estimates involve mathematical analysis, complex and nonlinear relationships among component parameters and performance metrics often require an iterative method. This is due to the limitations of existing mathematical models in fully capturing the intricate interactions between these elements. Therefore, designers rely on a trial-and-error approach, beginning with calculated size approximations and then refining them through extensive simulations. This iterative procedure includes the modification of transistor sizes based on the designer's experience and expertise to ensure that the overall circuit satisfies the specified performance requirement. Other components besides transistors, such as capacitors and resistors, are also included in the sizing step, resulting in complex and time-intensive problems to solve. Finally, the last step of the physical design phase involves the transformation of the optimized circuit diagram into a physical layout suitable for silicon chip fabrication. This layout process includes the placement of components on the floorplans, and routing these components using various connections of different materials and sizes. The final layouts are required to comply with predefined design regulations and guarantee manufacturability.

After each step and before manufacturing, strict verification schemes are implemented to ensure no flaw is found and regulations are not violated. Even after positive results, the products are susceptible to defection at a considerable probability. To accelerate and refine the analog IC design process and decrease human effort, numerous academic and commercial entities are actively researching various advanced methods to automate and optimize the circuit design process, especially in the component sizing phase. These methods, including bio-inspired optimization algorithms, quantum and quantum-inspired optimization algorithms, and advanced ML techniques, will be introduced and discussed in the subsequent chapters.

1.3 Analog Circuit Sizing Procedure

This section discusses the optimization process of analog circuit sizing. In what follows, the steps within this process are described in the order that enables a clear orientation of the procedure.

1.3.1 Topology Selection

Each analog circuit block has various options for its circuit topology. In case nominal topologies are available for selection, designers often carry out some initial analyses to match the specifications of the application, or else topology synthesis requires a much larger step to be executed as stated in previous sections. Even for smaller circuits, depending on the specific technical requirements to optimize one or more design parameters and their constraints, appropriate circuit configuration should still be considered thoroughly. This helps leverage the strength and mitigate the limitations of the chosen circuit topology to meet the predetermined technical requirements.

In short, although the initial stage of determining the optimal circuit topology may require a significant amount of time, it helps circuit designers obtain a clear direction to make the most suitable decisions. If more efforts are needed to synthesize the topology, designers are advised to finish it before going into the sizing procedure.

1.3.2 Parameter Selection for Optimization

Parameters for optimization in analog circuits are determined based on specific technical requirements. Along with that, it also requires identifying the design constraints of each parameter to ensure that the analog circuit optimization process does not prioritize the optimized parameters at the expense of others. Therefore, we aim for the maximum balance between all parameters in the circuit.

In other words, the optimized parameters would be enhanced to the fullest possible extent while the remaining parameters are simultaneously ensured to satisfy the design constraints.

1.3.3 Optimization Design Variables

To formulate an optimization problem for analog circuit sizing, the functions based on these parameters, such as the dimensions of the MOSFET channel (i.e., length (L) and width (W)), resistance, and capacitance, need to be established. Next, besides the parameters that need to be held constant, it is required to determine the variables that directly affect the design parameters to select appropriate optimization variables.

Since each technology node and its corresponding device have different ranges of values for the parameters mentioned above, their limits must be set. Furthermore, to possibly approach global optimum values with fewer iterations, it is crucial to limit the range of values of the optimization variables to narrow the search space, which increases the probability of improving the efficiency of the optimization process.

1.3.4 Objective Function and Design Constraints

Determining the objective function for the optimization problem of a similar circuit not only clarifies the purpose of the optimization problem but also serves as a crucial condition for implementing the optimization algorithm. The reason is that we need a specific value from the objective function to quantitatively evaluate a particular set of parameters for a similar circuit, thereby identifying the better set of parameters for the circuit. Furthermore, as mentioned above, to avoid imbalance among the parameters in the circuit, additional constraints on the parameters exclusive from the group of optimized parameters need to be imposed to achieve the maximum balance between those parameters.

1.4 Problem Modeling for Analog Circuit Sizing Optimization

In analog IC design tasks, the main goal is to determine the geometric metrics (width and length) of each transistor within the circuit to meet the predefined specifications. These specifications could include maximizing or minimizing specific performance metrics while ensuring that others exceed or remain below certain thresholds. By way of illustration, consider the design of a bandgap reference (BGR) circuit, which is employed to provide an accurate reference voltage for an entire mixed-signal chip. Various metrics are used to measure the performance of this circuit, such as power dissipation, temperature coefficient, power supply rejection ratio, and line sensitivity (which measures how the BGR output voltage changes in response to variations in the power supply voltage). In the context of sensor ICs, the primary concerns are power consumption—which should be low—and line sensitivity—which should be kept to a minimal [6]. In contrast, for BGRs intended for quantum computing circuits, its stability in a broad temperature range, particularly at cryogenic temperatures, takes precedence, making the temperature coefficient the most critical factor [7].

1.4.1 Problem Modeling for Analog Circuit Sizing Optimization

Determining the objective function and constraint functions for the optimization problem is dependent on the functionalities and the purpose of the circuit that we want to design and produce.

In general, the task of designing an analog IC can be formulated as the following optimization problem (1.1):

$$\arg\min_{\mathbf{x}} \quad \left[F_1(\mathbf{x}), F_2(\mathbf{x}), \dots, F_m(\mathbf{x})\right] \tag{1.1a}$$

$$\text{s.t.} \quad f_i(\mathbf{x}) \leq 0, \quad i = 1, 2, \dots, n, \tag{1.1b}$$

$$g_j(\mathbf{x}) = 0, \quad j = 1, 2, \dots, k, \tag{1.1c}$$

$$\mathbf{x} \in S, \tag{1.1d}$$

where:

- $\mathbf{x} \in \mathbb{R}$ represents the design variables, which include the width and length of transistors in the circuit, as well as resistance and capacitance values;
- $F_1(\mathbf{x}), F_2(\mathbf{x}), \dots, F_m(\mathbf{x})$ are the m functions that need to be optimized;
- $f_i(\mathbf{x})$ and $g_j(\mathbf{x})$ are unequal and equal constraints, representing the circuit specifications;
- $S \subset \mathbb{R}^+$ is the design space, limited by the upper and the lower bounds of each design variable.

When $m = 1$, the optimization problem is single-objective, while if $m > 1$, the problem is called a multi-objective optimization problem. In contrast to single-objective optimization, multi-objective optimization problems present a more nuanced challenge. In fact, obtaining a solution that can optimize all competing objective functions simultaneously is neither always possible nor straightforward. This is mainly due to the fact that these objectives are often in conflict: enhancing one may negatively affect the performance of others [8]. For instance, the design of a single-stage amplifier must take into account the trade-offs between several performance metrics, including power dissipation, supply voltage, linearity, noise, gain, speed, maximum voltage swings, input/output impedance [9]. For such a straightforward circuit, one can easily notice how intricate the conflict and support dynamics among its eight performance metrics are. This means that attempting to optimize more than one parameter, such as maximizing voltage swings and transition speed simultaneously, poses significant challenges. In fact, in many cases, achieving the maximum value for both parameters concurrently is very difficult, if not impossible, to achieve. To reach a satisfactory solution for both objective functions, various techniques can be employed. Among these, the more common techniques include (i) the weighted-sum technique which generates a new objective function by summing the original objective functions with weights, assigning a larger weighting coefficient to the more crucial objective function, (ii) the use of a utility function that generates a new objective function by multiplying or dividing the original objective functions with a suitable exponential degree for each, and (iii) the ϵ-constrained technique which retains only one objective function and transforms the other objective functions into constraints.

1.4.2 Criteria for Optimization Assessment

As mentioned in Section 1.2, state-of-the-art optimization techniques should be incorporated into the analog sizing procedure as a powerful engine to automate the optimal circuit design process, thus speeding up the overall analog IC design procedure and reducing human effort. In the realm of optimization, selecting the most suitable algorithm for a given problem is a critical task. Due to the lack of detailed understanding of the exact mechanism of these algorithms, for a specific problem, it is almost impossible to predict whether they will perform well without multiple attempts [10]. With a variety of optimization algorithms available, it becomes essential to employ robust criteria for comparison. Performance metrics serve as valuable tools in this regard, allowing systematic evaluation and comparison of the efficiency of different optimization algorithms. These metrics, providing quantitative insights into various aspects of the algorithm, can include:

- **Solution quality:** Evaluation of the quality of the solutions is instrumental in understanding how effectively an algorithm can identify optimal solutions that meet the objectives of the optimization problem. Superior-quality solutions, indicated by the better objective function values attained, are crucial to confirm that the algorithm solves the problem accurately and efficiently.
- **Convergence rate:** Convergence rate determines how quickly the algorithm can find an optimal or quasi-optimal solution. Faster convergence means fewer computational resources are required, which is particularly important for time-sensitive applications or large-scale optimization problems.
- **Robustness:** Optimization problems may exhibit variability or uncertainty. Robust algorithms can handle such variations and consistently produce good solutions, with respect to the optimal solution, the convergence speed, and the success rate. Robustness ensures that the algorithm's performance is reliable and not overly sensitive to minor changes in problem parameters.
- **Fair comparison:** Up to now, numerous optimization algorithms have been developed, and if considering their variants, there are more than a hundred distinct algorithms and versions. To select the most appropriate algorithm for a particular problem, it is essential to perform a fair comparison among them. However, the fairness of such a comparison depends on the criteria employed. A biased comparison may lead to unsuitable choices. Currently, there is no exhaustive literature that clearly defines a fair comparison methodology. The comparison criteria used in the current literature include precision, success rates, count of function evaluations, computational time, tolerance, mean values, optimal outcomes, and standard deviation. Whether these truly represent fair performance measures is debatable; thus, more comprehensive studies supported by rigorous mathematical analysis are urgently required.

1.4.3 Correlation Between Optimization Algorithms and Analog Circuit Sizing

Subsequent to determining the suitable optimization algorithm, it is necessary that the correlation between components of optimization algorithms and analog sizing be clarified. This should greatly help readers outline a general perspective of how optimization algorithms can be applied to analog sizing design. The optimization algorithm's entities can be interrelated as:

- **Population:** In optimization algorithms, a population demonstrating a set of candidate solutions is iteratively updated over the optimization process to find an optimal solution. When applied to analog sizing problems, the population includes the set of values for circuit components such as MOSFET's dimension, resistors, capacitors, inductors, and so on. Depending on the classification and characteristic of each algorithm, the population might be initialized differently, either in the form of decimal numbers, binary strings or quantum bits (qubits). In each case, a suitable method for conversion to compatible decimal values for circuit components' values needs to be applied. This is further clarified in the case studies presented in the subsequent chapters.
- **Objective function:** The objective function plays a vital role in an optimization algorithm as a means to assess the quality of each individual (or candidate solution) of the population. In analog circuit sizing, it is synonymous with the figure of merit (FoM), which is utilized to evaluate the performance of each combination of circuit components' values for a specific circuit.
- **Characteristic population update mechanism:** The update of individuals in the population can be considered algorithm-based, which mostly results from the algorithm's inspiration or specific characteristic mechanism. In analog sizing, circuit components, which are represented according to the algorithm's implementation, follow specific update methods prior to being converted to their compatible decimal format. When the algorithm employs decimal manipulation for its individuals, the circuit components do not require an additional conversion step.

References

1 G. Huang, J. Hu, Y. He, J. Liu, M. Ma, Z. Shen, J. Wu, Y. Xu, H. Zhang, K. Zhong, X. Ning, Y. Ma, H. Yang, B. Yu, H. Yang, and Y. Wang, "Machine Learning for Electronic Design Automation: A Survey," *ACM Trans. Des. Autom. Electron. Syst.*, vol. 26, no. 5, pp. 1–46, Jun. 2021.

2 D. Amuru, A. Zahra, H. V. Vudumula, P. K. Cherupally, S. R. Gurram, A. Ahmad, and Z. Abbas, "AI/ML Algorithms and Applications in VLSI Design and Technology," *Integration*, vol. 93, p. 102048, Jun. 2023.

3 P. P. Mercier, B. H. Calhoun, P.-H. P. Wang, A. Dissanayake, L. Zhang, D. A. Hall, and S. M. Bowers, "Low-Power RF Wake-Up Receivers: Analysis, Tradeoffs, and Design," *IEEE Open J. Solid-State Circuits Soc.*, vol. 2, pp. 144–164, Oct. 2022.

4 T. Q. Duong, L. D. Nguyen, B. Narottama, J. A. Ansere, D. V. Huynh, and H. Shin, "Quantum-Inspired Real-Time Optimization for 6G Networks: Opportunities, Challenges, and the Road Ahead," *IEEE Open J. Commun. Soc.*, vol. 3, pp. 1347–1359, Aug. 2022.

5 F. Zhang, *High-Speed Serial Buses in Embedded Systems*. Singapore: Springer, 2020.

6 B. Park, Y. Ji, and J.-Y. Sim, "A 490-pW SAR Temperature Sensor with a Leakage-Based Bandgap-Vth Reference," *IEEE Trans. Circuits Syst. II*, vol. 67, no. 9, pp. 1549–1553, 2020.

7 H. Homulle, F. Sebastiano, and E. Charbon, "Deep-Cryogenic Voltage References in 40-nm CMOS," *IEEE Solid-State Circuits Lett.*, vol. 1, no. 5, pp. 110–113, 2018.

8 X.-S. Yang, *Optimization Techniques and Applications with Examples*, 1st ed. Hoboken, NJ: Wiley, 2018.

9 B. Razavi, *Design of Analog CMOS Integrated Circuits*, 2nd ed. New York: McGraw-Hill Education, 2017.

10 X.-S. Yang, *Nature-Inspired Optimization Algorithms*. Academic Press, 2020.

2

Evolutionary Algorithms in Analog Circuit Sizing Optimization

In sizing optimization problems within integrated circuit (IC) design, it is often necessary to determine configurations that maximize circuit performance across a range of values for multiple parameter sets. Addressing this challenge requires efficient algorithms capable of exploring expansive solution spaces while ensuring convergence. Evolutionary principles, originally rooted in biological research, have emerged as a rigid foundation for developing optimization algorithms and have great potential to solve sizing optimization problems. Genetic algorithm (GA) and differential evolution (DE), two prominent evolutionary algorithms, leverage key biological processes such as crossover and mutation inspired by Charles Darwin's theory of evolution, to iteratively refine solutions and converge toward near-optimal configurations. Additionally, biogeography-based optimization (BBO), inspired by the natural distribution of species, models habitats based on environmental factors. This approach is analogous to representing circuit parameters in IC design as functions of circuit configurations. Building on these analogies and principles, this chapter explores the methodologies, and applications of GA, DE, and BBO in addressing the challenges of IC sizing optimization to effectively meet design specifications.

2.1 Genetic Algorithm

Genetic algorithm (GA), formulated by John Holland and his collaborators in the 1960s, belongs to the group of evolutionary computation methods employed for a wide variety of optimization problems [1]. Its application spans multiple fields, including graph coloring, pattern recognition, financial market analysis, and multi-objective discrete and continuous system engineering optimization [2].

Advanced Techniques for Optimal Sizing of Analog Integrated Circuits: Quantum Computing, Machine Learning, and Bio-inspired Optimization, First Edition.
Trang Hoang, Thinh Quang Do, Thang Quoc Nguyen, Hoang Trong Nguyen, Lihong Zhang, Octavia A. Dobre, and Trung Q. Duong.

GA is a directed random search technique based on Charles Darwin's principle of natural selection and biological evolution, in which suitable individuals best adapted for their environment survive and reproduce [3]. Driven by the concept of "survival of the fittest," better solutions are created throughout the generations' evolution thanks to the mechanics of natural selection, crossover, and mutation as a problem-solving strategy [4].

Compared to traditional optimization algorithms, GA is noticeably more advantageous with regard to its parallelism and capability to solve complex problems [5]. GA can handle multiple types of objective functions (or fitness functions) of optimization problems. These objective functions can be stationary or non-stationary, linear or nonlinear, continuous or discontinuous, or with random noise [2]. The mentioned flexibility in terms of objective function should be advantageous for analog sizing optimization, wherein objective functions can vary in different forms based on design specifications. Furthermore, GA's individuals play the role of independent agents, which enables the simultaneous search in different regions and different directions of the solution space, which might cater to the broad analog design space for optimal devices' size. Hence, this characteristic facilitates the parallel implementation of the algorithm.

Figure 2.1 and Algorithm 2.1 illustrate the implementation process of a GA, which will be clarified in Sections 2.1.1–2.1.6.

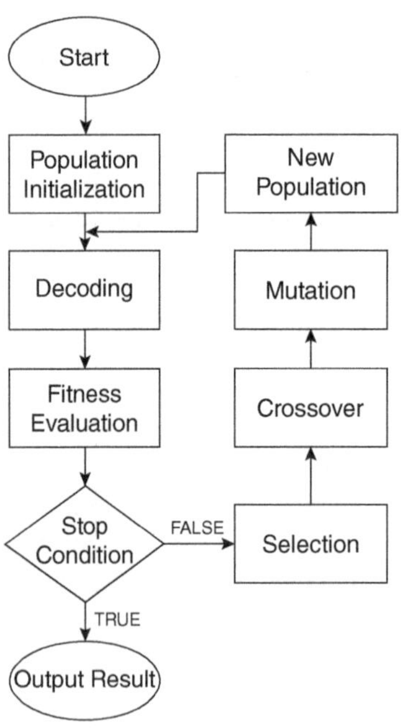

Figure 2.1 GA's flowchart. [5] / MDPI / CC BY 4.0.

Algorithm 2.1: Genetic Algorithm (GA) [5]

```
population = initial_population();
best = calculate_fitness(decode(population));
while !stop_condition do
    parents = select(population);
    children = crossover(parents);
    new_population = mutation(children);
    best = calculate_fitness(decode(new_population));
return decode(best);
```

2.1.1 Population Initialization

In Figure 2.1, a GA begins with the initialization of a fixed-size population with randomly created solutions to enable parallel search in the problem space. These solutions, typically encoded as fixed-length binary bit-strings, are modeled as chromosomes, which consist of genes encoded as either a single bit or a group of bits [5].

For a population of m n-bit chromosomes, the classical population has the general vector representation as

$$P_{GA} = \{(b_{11} \dots b_{j1} \dots b_{n1})_1, (b_{12} \dots b_{j2} \dots b_{n2})_2, \dots, \\ \times (b_{1i} \dots b_{ji} \dots b_{ni})_i, \dots, (b_{1m} \dots b_{jm} \dots b_{nm})_m\},$$

where $j \in \{1, 2, \dots, m\}, i \in \{1, 2, \dots, n\}$, and $b_{ji} \in \{0; 1\}$ (b_{ji} are binary bits).

2.1.2 Decoding

In every generation, the bit-strings of chromosomes' bit-strings are decoded from binary bit sequences to their decimal representations. Each binary bit sequence $(b_{1i} \dots b_{ji} \dots b_{ni})_i$ can be decoded into a decimal number to represent a value of the optimization variables.

The decoding step is given by

$$\text{value} = \text{LB} + \frac{\text{decimal}}{2^n} \times (\text{UB} - \text{LB}), \tag{2.1}$$

where value is the real number of the optimization variable; decimal is the unsigned decimal value of the binary sequence $(b_{1i} \dots b_{ji} \dots b_{ni})_i$; and LB and UB are the lower and upper bounds of the optimization variables, respectively.

2.1.3 Fitness Evaluation

The step of fitness evaluation aims to find the fitness values of all individuals in the population based on a predetermined fitness function (or objective function). Then, the fitness values of all individuals are assessed to identify the best individual in the current population, which also determines the margin of the current solutions with the predetermined target or the optimal solution of the problem [5].

The choice of fitness function can vary depending on the application or the desired performance metrics to be optimized. In every scenario, the fitness function holds a significant role as the standard for assessing the quality or "fitness" of potential solutions within the population. Consequently, the evolutionary process can be trained in a fitness-oriented manner.

2.1.4 Selection

The selection step in a GA involves choosing individuals from the current population to play the role of parents and form mating pools for subsequent generations. This step is typically driven by each individual's fitness, with fitter individuals being more likely to be selected. Various selection techniques such as tournament selection, roulette wheel selection, or rank-based selection can be used to determine which individuals are chosen as parents for reproduction.

The purpose of the selection step in GA is to guide the evolutionary process by favoring individuals with higher fitness values which will be chosen as parents for the next generation. By doing so, the algorithm promotes the propagation of traits that contribute to better solutions, enhancing the overall quality of the population over successive generations. In essence, selection ensures that the GA evolves toward better solutions by reflecting the process of natural selection in biological evolution.

2.1.5 Crossover

The crossover of two parent strings, which occurs with a high probability denoted as r_{cross} in the range [0.8; 0.95] [5, 6], is one of the main operators in GA. This step involves selecting individuals and exchanging one segment of one chromosome

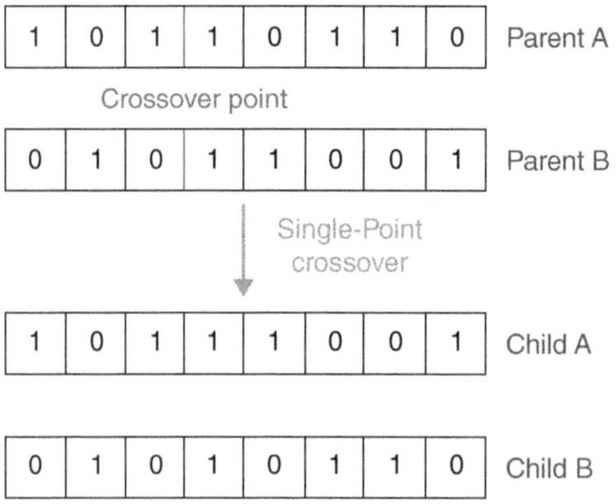

Figure 2.2 Single-point crossover at a random position.

with the corresponding segment on another chromosome at a random position to obtain new individuals. Figure 2.2 demonstrates the single-point crossover of two parent chromosomes, namely "Parent A" and "Parent B," to produce the results of "Child A" and "Child B."

The purpose of crossover in GA is to create offspring with traits inherited from both parent individuals. By exchanging genetic materials between parents, crossover facilitates the exploration of the solution space and the generation of potentially superior solutions. It stimulates diversity and introduces new combinations of traits, enabling the algorithm to move toward better solutions over successive generations.

2.1.6 Mutation

Mutation, which is usually implemented with low probability $r_m ut$ in the range [0.001; 0.05] [5, 6], is a genetic operator used to introduce random changes in individuals within the population. This step can be as straightforward as flipping a single bit in a bit-string, or it can involve more complex changes, such as complete genetic modifications [5, 7]. Mutation of a randomly selected bit is illustrated in Figure 2.3.

Its purpose is to maintain genetic diversity and prevent the algorithm from converging prematurely to suboptimal solutions. Through mutation, the algorithm can uncover new regions of the solution space that might not be reachable through crossover alone. It acts as a mechanism for introducing novel genetic material, potentially improving the overall search process and allowing the algorithm to escape local optima.

2.2 Self-Adaptive Differential Evolution

Differential evolution (DE), proposed by R. Storn and K. Price in 1997 [8], is a stochastic search algorithm with self-organizing and derivative-free characteristics [2]. Since both DE and GA are categorized into the evolutionary group, they utilize common operations, namely mutation and crossover, for generating new individuals and selection for determining the survival of these individuals for the next generation [2, 9]. However, DE is mainly different from GA with respect to its population representation. Instead of binary bits as in GA, DE uses real numbers as

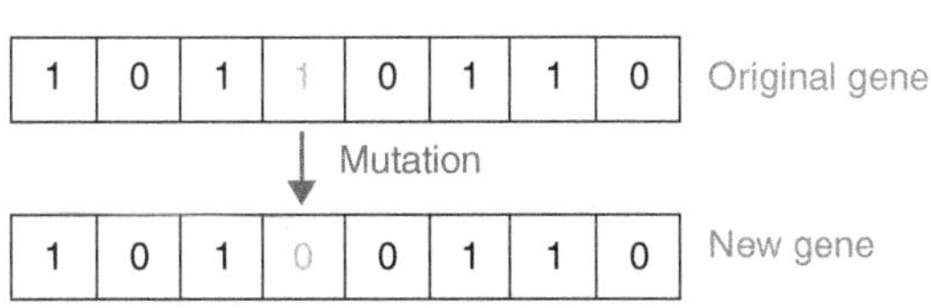

Figure 2.3 Mutation at a randomly selected bit.

solution strings, which does not require any further decoding step [2]. As a result, this decimal population representation can facilitate execution time and computational effort in the analog sizing optimization process, which mainly deals with device sizes to be optimized in real-number format.

DE can be implemented in several schemes based on the choice of mutation and crossover mechanism. The DE scheme in this section is classified as DE/Best/1/Bin [2, 8]. Moreover, the suitable choice for DE's control parameters is crucial to the performance of the optimization process. Nevertheless, the methodology for determining these control parameters is problem-dependent, experience-dependent, and inconsistent [9, 10], which might involve the time-intensive trial-and-error fine-tuning process. This is the primary motivation for proposing the self-adaptive DE (SaDE) algorithm, which includes an automatically altered parameter mechanism during the evolutionary process instead of fixed values to deal with different classes of problems. The mentioned mechanism is applied for F and CR, which are the differential weight and crossover rate, respectively. The pseudocode of the proposed approach is demonstrated in Algorithm 2.2 and clarified in Sections 2.2.1–2.2.4.

Algorithm 2.2: Self-Adaptive Differential Evolution (SaDE)

```
pop ⟵ Population Initialization(parameters);
fitness ⟵ Fitness Evaluation(pop);
while !stop_condition do
    pop_mutation ⟵ Mutation(pop);
    pop_crossover ⟵ Crossover(pop, pop_mutation);
    pop ⟵ Selection(pop, pop_crossover);
    fitness ⟵ Fitness Evaluation(pop);
return best solution;
```

2.2.1 Population Initialization and Fitness Evaluation

Initially, a population of m individual is generated; we have x_i where $i = 1, 2, \dots, m$. Thus, an m-individual population for SaDE, in which each individual has D dimensions, can be described as $P_{DE} = \{x_1, x_2, \dots, x_m\}$, where x_i denotes the decimal representation of the population.

Each solution x_i at any iteration t is given by $x_i^t = \{x_{i,1}^t, x_{i,2}^t, \dots, x_{i,D}^t\}$, where D is the dimension of the solution vector x_i.

The fitness values of the population are evaluated based on a fitness function or objective function of the optimization problem as

$$f_{obj}(x_i^t) = f_{obj}(x_{i,1}^t, x_{i,2}^t, \dots, x_{i,D}^t).$$

2.2.2 Mutation

For the implementation of the mutation step, firstly, two randomly distinct target vectors x_q, x_r are chosen at iteration t. Then, a donor vector v_i^{t+1} is calculated using the best-individual-based mutation scheme as follows:

$$v_i^{t+1} = x_{best}^t + F(x_q^t - x_r^t), \tag{2.2}$$

where x_{best}^t is the currently best individual until iteration t and F is the differential weight with real-number value in the range [0, 2]. However, the range of [0, 1] is more consistent and effective in practice [2].

In SaDE, rather than a fixed value, F is automatically updated according to

$$F = \begin{cases} F_l + rand \times F_u, & \text{if } rand < \tau_1, \\ F, & \text{otherwise}, \end{cases} \tag{2.3}$$

where F_l and F_u are the respective lower and upper bounds of F; *rand* is a uniform random value in the range [0, 1]; and τ_1 is the probability for adjusting F.

2.2.3 Crossover

Crossover is dependent on the so-called crossover parameter $CR \in [0, 1]$, which is the probability or the rate for the crossover operation. The SaDE algorithm utilizes the binomial crossover scheme, implemented on each of the D components of each individual according to

$$u_{i,j}^{t+1} = \begin{cases} v_{i,j}^t, & \text{if } r_j < CR, \\ x_{i,j}^t, & \text{otherwise}, \end{cases} \quad (j = 1, 2, \dots, D), \tag{2.4}$$

where $r_j \in [0, 1]$ is a uniformly distributed random number.

One of the contributions of the self-adaptive characteristic of SaDE involves the dynamic scheme for the CR parameter as

$$CR = \begin{cases} rand, & \text{if } rand < \tau_2, \\ CR, & \text{otherwise}, \end{cases} \tag{2.5}$$

where τ_2 is the probability for adjusting CR.

2.2.4 Selection

To select the fittest individuals for the next generation, the fitness-based selection is implemented as

$$x_i^{t+1} = \begin{cases} u_i^{t+1}, & \text{if } f_{obj}(u_i^{t+1}) \text{ is more optimal than } f_{obj}(x_i^t), \\ x_i^t, & \text{otherwise}. \end{cases} \tag{2.6}$$

2.3 Biogeography-Based Optimization

Biogeography-based optimization (BBO) is a newly proposed swarm heuristic optimization algorithm by Dan Simon in 2008 [11]. BBO has captured the attention of scholars and researchers worldwide thanks to its few parameters and straightforward, simple implementation [12]. The mechanism for BBO is inspired by the science of biogeography, a study of the natural distribution of biodiversity over space and time. This section begins by exploring the key concepts and theories of biogeography, and then proceeds to apply them in the formulation of BBO.

2.3.1 Biogeography: The Science of Evolution

The science of biogeography was introduced by two naturalists named Alfred Wallace [13] and Charles Darwin [14]. Until 1967, biogeography was primarily a descriptive and historical field. However, Robert H. MacArthur and Edward O. Wilson revolutionized the field by developing a mathematical model for biogeography [15], enabling predictions about the number of species in a given habitat. In this model, a habitat, defined as an ecological area occupied by a specific plant or animal species geographically isolated from other habitats, possesses a certain level of favorability classified by the habitat suitability index (HSI). Habitats that are highly suitable as living environments for biological species are characterized by a high HSI and vice versa. The HSI value depends on various factors, including rainfall, temperature, humidity, illumination, vegetation diversity, land area, safety and so on. These factors, known as suitability index variables (SIVs), play the role of independent variables. By assigning values to each SIV, the HSI can be expressed as a dependent variable in the form of a function of these SIV inputs. Analogously, in analog circuit design, a circuit parameter may be dependent on a few or several circuit elements. For instance, the voltage gain of an operational amplifier can be expressed as a function of the MOSFETs' transconductance and transresistance, which might also be determined by these MOSFETs' channel width, channel length, and several other physical constant parameters.

Habitats with a high HSI typically support a large number of species, while those with a low-HSI host fewer species. High-HSI habitats exhibit a high emigration rate μ due to the abundance of species, creating more opportunities for species to emigrate to neighboring areas. However, their immigration rate λ is low, as these habitats are often close to being fully occupied by species. Consequently, high-HSI habitats tend to have stable species distributions. On the other hand, low-HSI habitats have a high immigration rate λ and a low emigration rate μ owing to their low population density. The arrival of new species to low-HSI habitats can increase the habitat's HSI since biological diversity enhances their suitability. By the same token, in analog sizing optimization, when a specific circuit parameter needs improving, circuit elements' modification can be directed toward those with the highest impact on this parameter, possibly at an acceptable expense of other parameters. However, if the habitat's HSI remains low, its species are more likely to go extinct, creating opportunities for further immigration. This cycle makes low-HSI habitats more dynamic and subject to frequent changes in species composition compared to the more stable high-HSI habitats.

For simplicity, it is reasonable to assume a linear relationship between HSI and its uniform immigration and emigration rates across all habitats. The immigration and emigration rates are directly influenced by the number of species in the habitat. These relationships are illustrated in Figure 2.4, with the immigration rate λ and emigration rate μ expressed as

$$\lambda = I\left(1 - \frac{k}{n}\right), \tag{2.7}$$

$$\mu = \frac{E}{n}, \tag{2.8}$$

where I is the maximum possible immigration rate, E is the maximum possible emigration rate, k is the number of species of the kth individual, and n is the number of species.

When $E = I$, combining (2.7) and (2.8), we obtain

$$\mu + \lambda = E. \tag{2.9}$$

According to Figure 2.4, the immigration rate λ and emigration rate μ of the habitat are influenced by its species population. The immigration rate is highest with its maximum value of I when the habitat has zero species, as there is no

Figure 2.4 Habitat migration rate (including μ and λ) versus the number of species in a habitat.

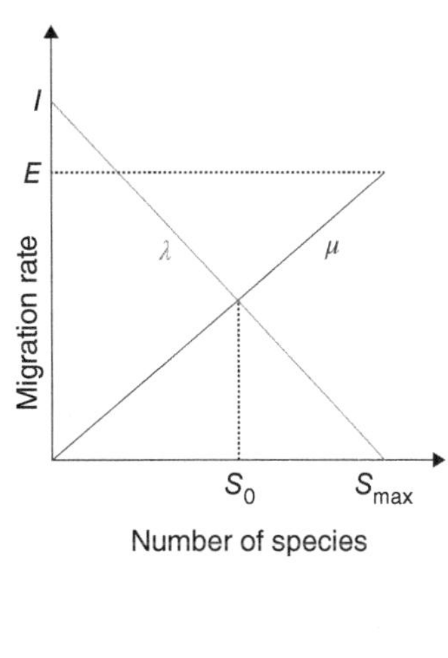

competition for resources. However, with the rise in the number of species, the habitat becomes more densely populated, reducing the likelihood of successful immigration. When the habitat reaches its maximum capacity S_{max}, the immigration rate drops to zero. Conversely, the emigration rate is zero when no species are present in the habitat. As the population grows and the habitat becomes more crowded, the likelihood of species leaving to explore other areas increases. The emigration rate reaches its maximum E when the habitat supports its largest possible number of species, S_{max}.

Furthermore, it can be observed from Figure 2.4 that the equilibrium number of species, S_0, is reached when the immigration rate and emigration rate are equal, or when the curve of λ crosses that of μ. However, temporary deviations, including positive and negative deviations, from S_0 may occur due to various factors. On the one hand, positive deviations might result from a sudden surge in immigration, such as the arrival of a large piece of flotsam from a neighboring habitat, or a rapid burst of speciation akin to a miniature Cambrian explosion. On the other hand, negative deviations could arise from events like disease outbreaks, newly introduced voracious predators, or other natural disasters. Subsequent to significant disruptions, it may take considerable time for species counts to return to their equilibrium state [16].

The concepts of immigration and emigration can be mathematically represented using a probabilistic model. Let P_s denote the probability that the habitat contains exactly S species at time t. This probability is given by

$$P_s(t + \triangle t) = P_s(t)(1 - \lambda_s \triangle t - \mu_s \triangle t) + P_{s-1}\lambda_{s-1} \triangle t + P_{s+1}\lambda_{s+1} \triangle t, \tag{2.10}$$

where λ_s and μ_s are the immigration and emigration rates when there are S species in the habitat. This equation is valid since in order to have S species at time $t + \triangle t$, one of the following conditions must hold [11]:

- There were S species at time t, and no immigration or emigration occurred between t and $t + \triangle t$
- There were $S - 1$ species at time t, and one species immigrated
- There were $S + 1$ species at time t, and one species emigrated.

If the time interval $\triangle t$ is sufficiently small, the probability of more than one immigration or emigration within this period can be neglected. By taking the limit as $\triangle t \to 0$, (2.10) can be simplified to

$$P_s = \begin{cases} -(\lambda_s + \mu_s)P_s + \mu_{s+1}P_{s+1} & \text{if } S = 0, \\ -(\lambda_s + \mu_s)P_s + \lambda_{s-1}P_{s-1} + \mu_{s+1}P_{s+1} & \text{if } 1 \leq S \leq S_{max} - 1, \\ -(\lambda_s + \mu_s)P_s + \lambda_{s-1}P_{s-1} & \text{if } S = S_{max}. \end{cases} \tag{2.11}$$

2.3.2 Biogeography and Optimization

BBO is a newly developed optimization algorithm that draws inspiration from biogeography. Therefore, based on the concepts and theories of biogeography, this section demonstrates its application to formulate BBO regarding its two main operators, namely migration and mutation.

2.3.2.1 Migration

In the BBO algorithm, a population of candidate solutions is represented in the form of real-number vectors, with each number corresponding to an SIV. These SIVs are used to evaluate the fitness of each candidate solution, represented by its HSI value. Solutions with higher HSI values indicate better quality, while those with lower HSI values represent poorer-quality solutions.

In the BBO algorithm, solutions, representing habitats and employing emigration and immigration rates, are used to probabilistically exchange information between them [11]. Each solution has a probability, called the habitat modification probability P_{mod}, of being modified based on others. When a solution S_i is chosen to be modified, its immigration rate λ determines whether each SIV should be updated. For a selected SIV of S_i, the emigration rates μ of other solutions are used to choose which habitat will provide a randomly selected SIV to replace it [12]. This modification process might be suitable for analog optimization, which involves vectors of optimization variables in the form of circuit elements. To preserve the quality of the best solutions, BBO incorporates an elitism operation, ensuring that the top habitats, those with the highest HSI, remain unchanged during migration after each iteration.

The migration operator is illustrated in Algorithm 2.3.

Algorithm 2.3: BBO's Migration

Select a habitat H_i with probability $\propto \lambda_i$;
if *H_i is selected* **then**
    for *$j = 1 : n$* **do**
        Select H_j with probability $\propto \mu_i$;
        if *H_j is selected* **then**
            Randomly select an SIV γ from H_j;
            Replace a random SIV in H_j with γ;

2.3.2.2 Mutation

In nature, events like natural calamities can abruptly alter a habitat's HSI. In the BBO algorithm, such events are modeled as mutations in SIVs, with mutation rates determined by species count probabilities. These probabilities are derived using the differential equation from (2.11) and reflect the likelihood of each habitat existing as a solution to the problem. Solutions with very low probabilities are more likely to mutate into new solutions, while those with high probabilities have a lower chance of mutating. Consequently, solutions with very high or very low-HSI values are less likely to generate significant improvements in their SIVs during later stages. In contrast, medium-HSI solutions have a greater potential to evolve into better solutions through mutation. Thanks to this mechanism, exploration and exploitation can be balanced to the greatest possible extent, which should be beneficial for analog sizing optimization with a large, multi-dimensional and highly complex design space. The mutation rate $m(S)$ for each solution set is calculated based on species count probabilities as

$$m(S) = m_{max}\left(\frac{1 - P_S}{P_{max}}\right), \tag{2.12}$$

where m_{max} is a user-defined parameter.

The mutation scheme in BBO is designed to promote diversity within the population. Without it, highly probable solutions would dominate, reducing variation. Mutation increases the likelihood of modification for low-HSI solutions, offering them opportunities for improvement, and for high-HSI solutions, enabling further enhancement. To safeguard the best solutions, an elitist approach is used, preserving their features even if mutation ruins their HSI, allowing recovery if necessary. While the highly risky mutation process is applied to both poor and strong solutions, average solutions are typically left unchanged to encourage their natural improvement. However, there remains a small mutation probability for all solutions, except for the most probable ones.

BBO's mutation operator can also be demonstrated using Algorithm 2.4.

Algorithm 2.4: BBO's Mutation

```
for j = 1: length(SIV) do
    Compute P_i using λ_i, μ_i;
    Select SIV H_i(j) with probability ∝ P_i;
    if H_i(j) is selected then
        Replace H_i(j) with a randomly generated SIV;
```

2.4 Case Study: Single-Tail Dynamic Comparator[1]

Comparators are considered the heart of analog-to-digital converters (ADCs). They are used as a means to convert analog domain signals to digital domain

1 This section has been published as part of [5].

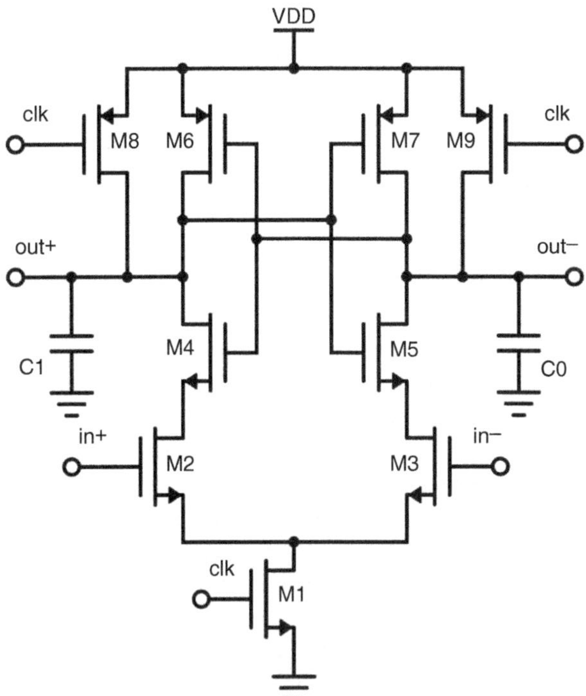

Figure 2.5 Schematic of the conventional single-tail dynamic comparator.

signals in modern signal processing and communications. In the design of high-speed ADCs, low-power and high-speed comparators are in great demand [17]. Thanks to strong positive feedback and dynamic bias provided by a pair of cross-coupled inverters as the latching stage, dynamic comparators have higher speed and less static power consumption compared to static comparators [18]. Therefore, with a view to optimizing comparators' performance regarding speed and power consumption, the dynamic comparator is chosen as a feasible candidate.

Section 2.4.1 illustrate the proposed Python–Spectre framework for optimizing the dynamic comparator circuit (depicted in Figure 2.5), followed by its setup steps for the GA-based optimization process as well as a discussion of the optimization results. Further analysis for the conventional dynamic comparator can be found in [5].

2.4.1 Proposed Python–Spectre Model to Optimize Delay and Power of the Dynamic Comparator

When applying GA to optimization problems, the structure of the algorithm almost remains unchanged. Meanwhile, it is necessary to make suitable modifications to the fitness calculation step for specific problems. For the optimization of analog circuit design, the "Fitness Evaluation" block in Figure 2.1 consists of two substeps. The first one is delay and power simulations, which are executed by the Cadence Virtuoso tool in the TSMCN65 process, with the help of the Spectre simulation platform. The remaining step is the evaluation of fitness values according to delay and power data collected in the previous step.

In this subsection, GA is implemented by the Python programming language since Python is currently considered the most widely used programming language and has various built-in libraries for AI, ML, and quantum machine learning (QML) algorithms' programming. In addition, Spectre is chosen as the circuit simulator for the whole design as an alternative to related studies. Circuit simulations in the works of [19, 20] were performed by HSPICE, whose output data were mostly presented in the simulator's predetermined format. Thus, scripting languages are usually necessary to create files containing output results of compatible format for Python processing of GA. On the other hand, it is possible for SKILL programming's syntax to be integrated into Ocean-based scripts of the Spectre simulator. This means that simulation results can be arranged in users' desirable format without further use of scripting languages. On account of its function in the optimization system and convenience in output data format, Spectre is preferably chosen as a practical candidate for circuit simulations.

The interrelation between Spectre and Python is described as follows. At first, the values for the population of design variables are created by Python and sent to Spectre via an Ocean-based script. This Ocean script is responsible for automated circuit simulations based on received numbers, and the results are sent back and further processed by the Python script to evaluate the fitness scores of each individual. This repetitive process carries on until the algorithm reaches its stopping point. The mentioned Python–Spectre correlation is further clarified in Figure 2.6.

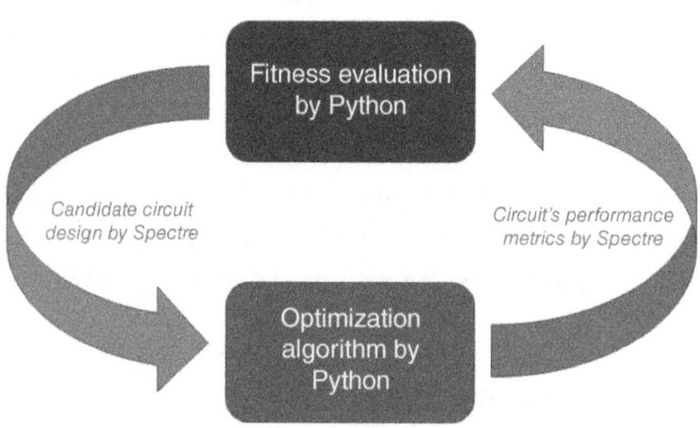

Figure 2.6 Mutual interaction between Python and Spectre platforms. [5] / MDPI / CC BY 4.0.

When optimizing the delay and power of the dynamic comparator, first and foremost, determining the optimization variables—parameters that mainly contribute to delay and power results—is essential. Since we use the TSMCN65 process for the design, the lengths of all MOSFETs in the circuit are set to 65nm. According to the delay analysis, since the difference between input voltages $\triangle V_{in}$ of the dynamic comparator is fixed at 5mV, the current through MOSFET M_1 (I_1) and capacitors C_i ($i = \{0, 1\}$) is responsible for delay adjustment (notice that M_i denotes the ith MOSFET of the dynamic comparator circuit). Therefore, we declare two delay-related variables: the width of $M_1(W_1)$ and the capacitors' values (C_i). Similarly, as the values of the supply voltage V_{DD} and the frequency of the clock signal *clk* are assigned to 1.2V and 1GHz, respectively, our power-related variables are determined based on the current I_{supply}, which is the sum of four currents flowing through M_6, M_7, M_8, and M_9. Due to the symmetry of the circuit, the widths of M_6 and M_7, and M_8 and M_9 should be equal. Hence, we declare two more variables, namely the width of M_6 and M_7 (W_{67}) and the width of M_8 and M_9 (W_{89}). For the remaining MOSFETs, the widths of M_2 (W_2) and M_3 (W_3), and M_4 (W_4) and M_5 (W_5) are also set to equal values to ensure the circuit's symmetry. For optimal performance of delay and power consumption, the simulation results indicate that $W_2 = W_3 = 0.21\,\mu m$, and $W_4 = W_5 = 0.12\,\mu m$. In total, we need four optimization variables, namely W_1, W_{67}, W_{89}, and C_i. To satisfy the range of the process and assure the functional correctness of the conventional dynamic comparator, the simulation results indicate that the bounds for the four above-mentioned variables are [0.12 μm; 2 μm], [0.12 μm; 2 μm], [0.12 μm; 2 μm], and $[0.1\,fF; 0.8\,fF]$, respectively.

For subsequent GA initializing steps, the number of bits for each chromosome representation is defined as 16 bits, and the population size is chosen to be six individuals per population. While crossover has a high probability, the probability of mutation is typically low; the authors in [3] showed that r_{cross} and r_{mut} are inside the ranges [0.8; 0.95] and [0.001; 0.05], respectively. Finally, the choice of a suitable fitness function is also critical. The figure of merit (FoM) to compare the performance of dynamic comparators can be either energy efficiency as in [21] or power-delay product (PDP) as in [22]. With a view to optimizing both the delay and power of the dynamic comparator, we choose the FoM in the form of PDP as the fitness function for GA. The comparator is considered to perform better with a lower value of PDP, which means that a lower PDP is equivalent to higher fitness values.

In other words, our optimization problem can be summarized as

$$\textit{minimize} \quad \mathrm{PDP}(W_1, W_{67}, W_{89}, C_i), \tag{2.13a}$$

$$\textit{s.t.} \quad L = 65\,\mathrm{nm}, \;\; W_2 = W_3 = 210\,\mathrm{nm}, \;\; W_4 = W_5 = 120\,\mathrm{nm}, \tag{2.13b}$$

$$\triangle V_{in} = 5\,\mathrm{mV}, \;\; V_{DD} = 1.2\,\mathrm{V}, \;\; f_{clk} = 1\,\mathrm{GHz}, \tag{2.13c}$$

$$0.12\,\mu\text{m} \leq W_1, W_{67}, W_{89} \leq 2\,\mu\text{m}, \tag{2.13d}$$

$$0.1\,fF \leq C_i \leq 0.8\,fF. \tag{2.13e}$$

2.4.2 Results and Discussion

The simulation results indicate that the lowest value for PDP of 0.2254 fJ is achieved for the case of $r_{cross} = 0.8$ and $r_{mut} = 0.05$. With $PDP = 0.2254\,fJ$, the delay and power of the conventional dynamic comparator are 72.48ps and 3.11 μW, respectively. PDP fitness values, as well as the delay and power over 100 iterations of GA, are illustrated in Figures 2.7 and 2.8, respectively.

As can be clearly observed in Figure 2.8, the values for delay and power vary in an unpredictable and non-monotonous manner. However, their corresponding PDP in Figure 2.7 decreases monotonously throughout 100 iterations. Since GA produces chromosomes with better fitness values at the end of each iteration, PDP's downward trend conforms to the working principle of the algorithm. In terms of the variables declared for GA, the optimal set $(W_1, W_{67}, W_{89}, C_i) = (0.4463, 0.1277, 0.1553\,\mu\text{m}, 0.1\,fF)$ is obtained after 100 iterations of the algorithm. Nevertheless, it is worth noticing that the process grid of the TSMCN65 process is 5nm. This means that the widths of

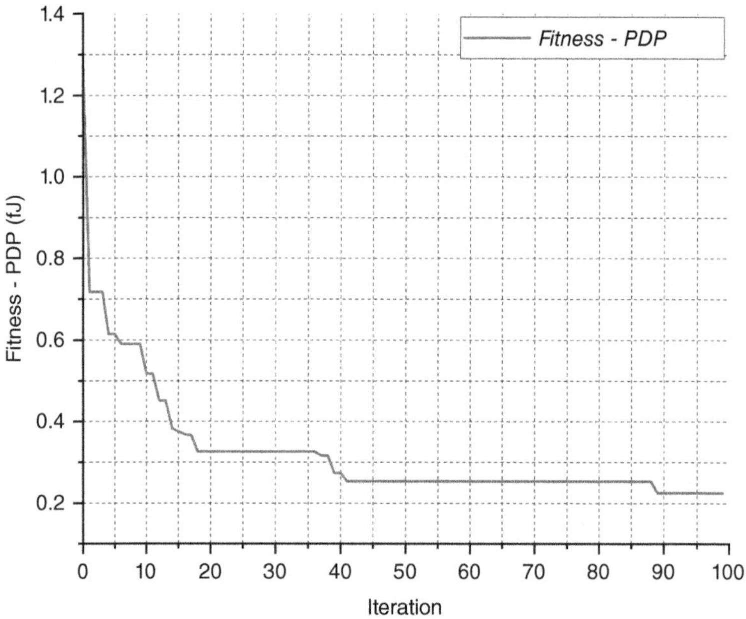

Figure 2.7 Fitness value of PDP versus 100 iterations.

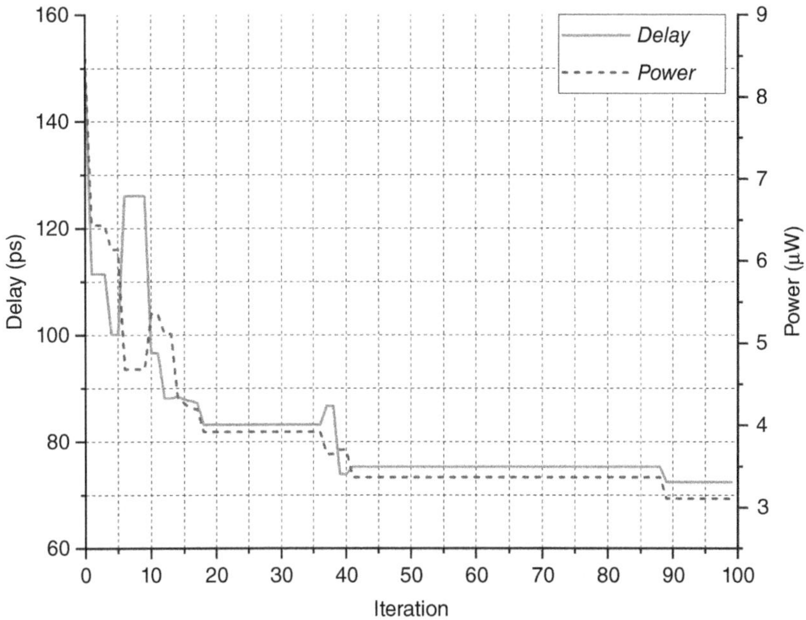

Figure 2.8 Delay and power versus 100 iterations.

M_1, M_6, M_7, M_8, M_9 need to be rounded to their closest feasible values as $(W_1, W_{67}, W_{89}) = (0.445, 0.13, 0.155\,\mu\text{m})$. Re-simulated results with the set $(W_1, W_{67}, W_{89}, C_i) = (0.445, 0.13, 0.155\,\mu\text{m}, 0.1\,fF)$ vary slightly with the final values of 72.61 ps for delay, 3.11 μW for power consumption, and 0.2258 *fJ* for PDP. Table 2.1 presents the post-optimization sizes of all MOSFETs in the circuit, while Table 2.2 summarizes the performance of the conventional dynamic comparator considered in this section against those in other studies.

Table 2.1 Post-optimization transistors' sizes.

Device	Size (W/L)
M_1	0.445 μm/65 nm
M_2, M_3	0.21 μm/65 nm
M_4, M_5	0.12 μm/65 nm
M_6, M_7	0.13 μm/65 nm
M_8, M_9	0.155 μm/65 nm
C_0, C_1	0.1 *fF*

Table 2.2 Performance summary of different dynamic comparator designs.

Parameters	[17]	[21]	[23]	[24]	[25]	This work
CMOS process (nm)	90	65	65	90	40	65
Supply voltage (V)	1.2	1	1.2	1	1.1	1.2
Clock frequency (GHz)	3.07	20	6	1	6	1
Average delay (ps)	410	14.28	42.7	51.76	54	72.61
Power dissipation (μW)	0.24	67.8	381	32.62	288	3.11
PDP (*fJ*)	0.0984	0.968	16.3	1.67	15.552	0.2258
Energy per conversion (*fJ*/conv.)	0.07818	3.39	63.5	32.62	48	3.11
Kickback noise (mV)	N/A	N/A	N/A	N/A	N/A	122

As the conventional dynamic comparator in this section has zero static power consumption, its power consumption at 1 GHz is much lower than that of the circuit in [24]. Since clock frequency is directly proportional to power dissipation, the designs of [21, 23, 25] with higher clock frequency exhibit a higher power than our design, which is reasonable. Meanwhile, the power consumption in [17] is still lower despite operating at a higher frequency. The parameter energy per conversion, which is equal to the ratio of power oversampling frequency (or clock frequency), is therefore needed to evaluate dynamic comparators' performance with respect to power. From Table 2.2, it is clear that our proposed design has the second-lowest energy per conversion value at 3.11 fJ per conversion.

In addition, compared to [23–25], our work has approximately a 20% higher average delay. On the contrary, our average delay is less than one-fifth in comparison with the delay of [17]. Because delay and power trade-off with each other, PDP is utilized as the FoM when optimizing both parameters. With respect to PDP, our design obtains the second-best value of 0.2258*fJ* versus the lowest number of 0.0984*fJ* by [17].

For further assessment of our optimization system, the optimization platform of the Analog Design Environment GXL, which offers both local and global optimization of circuit performances of interest, can be utilized as a suitable reference model. For setup steps, PDP is chosen as the optimization function, and the optimization variables and their bounds are similar to the GA-Spectre-based optimization system. Regarding ADE GXL's local optimization, the post-optimization transistors' sizes $(W_1, W_{67}, W_{89}, C_i) = (0.445, 0.13, 0.155\,\mu\text{m}, 0.1\,fF)$ obtained from the GA-Spectre framework are set as the reference points. With regard to global optimization from ADE GXL, due to our constrained data, the C version Feasible Sequential Quadratic Programming (CFSQP) is selected by the Spectre

Table 2.3 Comparison between different optimization models.

Parameters	ADE GXL's Local Optimization	ADE GXL's Global Optimization	Our GA-based Optimization
PDP (*fJ*)	0.227	0.236	0.2258

simulator as the optimization algorithm. It is worth acknowledging that while the CFSQP utilizes a single individual for each iteration rather than a population of individuals, GA in our system is implemented on a population of six individuals per iteration. Hence, for a decent comparison, the PDP result of 100 iterations of our GA-based optimization system should be compared with that of $6 \times 100 = 600$ iterations of the ADE GXL's global optimization tool.

Table 2.3 indicates that the result of PDP acquired by our optimization system is lower and has a higher convergence rate compared to that of ADE GXL's local as well as global optimization tool.

References

1 J. H. Holland, "Genetic Algorithms," *Sci. Am.*, vol. 267, no. 1, pp. 66–73, Jul. 1992.

2 X.-S. Yang, *Optimization Techniques and Applications with Examples.* New Jersey: Wiley, 2018.

3 X.-S. Yang, S. F. Chien, and T. O. Ting, "Chapter 1 - Bio-inspired computation and optimization: An overview," in *Bio-Inspired Computation in Telecommunications*, X.-S. Yang, S. F. Chien, and T. O. Ting, Eds. Boston, MA, England: Morgan Kaufmann, 2015, pp. 1–21.

4 D. E. Goldberg, *Genetic Algorithms.* Chennai, India: Pearson Education India, 2013.

5 H. T. Nguyen and T. Hoang, "A Novel Framework of Genetic Algorithm and Spectre to Optimize Delay and Power Consumption in Designing Dynamic Comparators," *Electronics*, vol. 12, no. 16, p. 3392, Aug. 2023.

6 F. Rothlauf, *Optimization Methods.* Berlin, Heidelberg: Springer Berlin Heidelberg, 2011, pp. 45–102.

7 B. Liu, Y. Wang, Z. Yu, L. Liu, M. Li, Z. Wang, J. Lu, and F. V. Fernández, "Analog Circuit Optimization System Based on Hybrid Evolutionary Algorithms," *Integration*, vol. 42, no. 2, pp. 137–148, Feb. 2009.

8 R. Storn and K. Price, "Differential Evolution–A Simple and Efficient Heuristic for Global Optimization Over Continuous Spaces," *J. Global Optim.*, vol. 11, pp. 341–359, Jan. 1997.

9 J. Brest, S. Greiner, B. Boskovic, M. Mernik, and V. Zumer, "Self-Adapting Control Parameters in Differential Evolution: A Comparative Study on Numerical Benchmark Problems," *IEEE Trans. Evol. Comput.*, vol. 10, no. 6, pp. 646–657, Dec. 2006.

10 M. H. Maruo, H. S. Lopes, and M. R. Delgado, *Self-Adapting Evolutionary Parameters: Encoding Aspects for Combinatorial Optimization Problems*. Berlin, Heidelberg: Springer Berlin Heidelberg, 2005, pp. 155–158.

11 D. Simon, "Biogeography-Based Optimization," *IEEE Trans. Evol. Comput.*, vol. 12, no. 6, pp. 702–713, Dec. 2008.

12 Z. Zhang, Y. Gao, and W. Zuo, "A Dual Biogeography-Based Optimization Algorithm for Solving High-Dimensional Global Optimization Problems and Engineering Design Problems," *IEEE Access*, vol. 10, pp. 55988–56016, May 2022.

13 A. R. Wallace, *The Geographical Distribution of Animals*. Cambridge, England: Cambridge University Press, 1876.

14 C. Darwin, *The Origin of Species*. Cambridge, England: Cambridge University Press, 2009.

15 R. H. MacArthur and E. O. Wilson, *The Theory of Island Biogeography*, Princeton Landmarks in Biology. New Jersey: Princeton University Press, 2001.

16 I. Hanski and O. E. Gaggiotti, *Metapopulation Biology: Ecology, Genetics, and Evolution*. Ontario, Canada: Academic Press, 2004.

17 S. Jakhar, V. S. Mandloi, R. Goswami, and K. Kandpal, "A Low Power, High Speed 1.2 V Dynamic Comparator for Analog-to-Digital Converters," *Procedia Comput. Sci.*, vol. 171, pp. 1018–1026, Jan. 2020.

18 Y. Wang, M. Yao, B. Guo, Z. Wu, W. Fan, and J. J. Liou, "A Low-Power High-Speed Dynamic Comparator with a Transconductance-Enhanced Latching Stage," *IEEE Access*, vol. 7, pp. 93396–93403, Jul. 2019.

19 S. R. Chowdhury, S. Bhardwaj, and J. Kitchen, "Design automation of CMOS op-amps using statistical geometric programming," in *Proc. IEEE Int. Symp. Circuits Syst. (ISCAS)*, Austin, TX, USA, May 2022, pp. 1575–1579.

20 M. V. Harsha and B. P. Harish, "An integrated MaxFit genetic algorithm-SPICE framework for 2-stage op-amp design automation," in *Proc. IEEE Comput. Soc. Ann. Symp. VLSI (ISVLSI)*, Hong Kong, China, Jul. 2018, pp. 170–174.

21 J. K. Folla, M. Crespo, E. T. Wembe, M. Bhuiyan, A. Cicuttin, B. Essimbi, and M. B. I. Reaz, "A Low-Offset Low-Power and High-Speed Dynamic Latch Comparator with a Preamplifier-Enhanced Stage," *IET Circuits Devices Syst.*, vol. 15, no. 1, pp. 65–77, Dec. 2020.

22 L. Samid, P. Volz, and Y. Manoli, "A dynamic analysis of a latched CMOS comparator," in *Proc. 2004 IEEE Int. Symp. Circuits Syst. (ISCAS)*, Vancouver, BC, Canada, Sep. 2004, pp. I–181.

23 H. Ghasemian, R. Ghasemi, E. Abiri, and M. R. Salehi, "A Novel High-Speed Low-Power Dynamic Comparator with Complementary Differential Input in 65nm CMOS Technology," *Microelectron. J.*, vol. 92, p. 104603, Oct. 2019.

24 V. Savani and N. M. Devashrayee, "Analysis and Design of Low-Voltage Low-Power High-Speed Double Tail Current Dynamic Latch Comparator," *Analog Integr. Circuits Signal Process*, vol. 93, pp. 287–298, Sep. 2017.

25 S. Huang, L. He, Y.-K. Chou, and F. Lin, "A 288-μW 6-GHz hybrid dynamic comparator with 54-ps delay in 40-nm CMOS," in *Proc. 2016 IEEE MTT-S Int. Wirel. Symp. (IWS)*, Shanghai, China, Mar. 2016, pp. 1–4.

3

Animal-Behavior-Inspired Algorithms in Analog Circuit Sizing Optimization

This chapter introduces various optimization algorithms inspired by both fauna and flora behaviors, including reproduction, foraging, preying, and the balance between exploration and exploitation. Particle swarm optimization (PSO) belongs to the group of foraging- and hunting-inspired algorithms, drawing from the foraging behavior of birds toward positions with the most abundant food. Similarly, ant colony optimization (ACO) simulates the process of ant foraging behavior and pheromone-based communication, guiding the search for optimal solutions. Bat algorithm (BA) mimics echolocation used by bats for prey detection. In the reproduction-inspired group, the firefly algorithm (FA) is based on attraction behaviors driven by brightness and intensity, while cuckoo search (CS) draws inspiration from cuckoos' parasitic reproductive strategies. Flower pollination algorithm (FPA) simulates the natural process of pollination, where pollinators transfer pollen between flowers, enabling reproduction. Most of these algorithms exhibit underlying characteristics of exploration and exploitation processes based on individual or group experiences in choosing the best individuals or strategies to reach their targets. This analogy to IC designs highlights their potential application, further clarified through a case study of implementing the CS algorithm in BGR circuit design.

3.1 Particle Swarm Optimization

Particle swarm optimization (PSO) is one of the algorithms constructed to search for solutions to optimization problems in a given search space. It is a type of population-based algorithm, inspired by the group behavior of birds when searching for food: during the foraging process, each bird with predetermined inertia tends to adjust its trajectory toward the location with the highest amount of food it has previously discovered, while also aiming toward positions with the most

Advanced Techniques for Optimal Sizing of Analog Integrated Circuits: Quantum Computing, Machine Learning, and Bio-inspired Optimization, First Edition.
Trang Hoang, Thinh Quang Do, Thang Quoc Nguyen, Hoang Trong Nguyen, Lihong Zhang, Octavia A. Dobre, and Trung Q. Duong.

abundant food that the others in the group have found so far. The PSO algorithm has many vital applications in various fields that require solving optimization problems.

Algorithm 3.1: Standard PSO Algorithm's Pseudo-code

Initialize a swarm of N particles;
for not stopping **do**
  fitness[i] = fitness_calculation()
  Pbest[i] = (fitness[i] < Pbest_val[i]) * fitness[i] + (fitness[i] > Pbest_val[i]) * Pbest[i]
  $Gbest = \min_{1 \le i \le N} (Pbest[i])$
  position[i], velocity[i] = update()
return Gbest's position and fitness value

Algorithm 3.1 summarizes the standard PSO algorithm. In this algorithm, the first step is to randomly initialize a set of N different coordinates for the design variable vector. In PSO, this set is called **swarm**, and each coordinate is called a **particle**. Later, the fitness function's value of each particle is calculated from the position, and then, it is evaluated: the newly calculated fitness value is compared with the fitness value of *Pbest* (the best fitness value that a particle has found so far). Suppose that the current fitness value of the particle is smaller than the present value of Pbest's fitness function. In that case, the current fitness value and coordinate are recorded as the new Pbest's fitness value and coordinate. Following this, if there exists any particle in the swarm whose Pbest's fitness is smaller than that of the current *Gbest* (the best position found so far by the entire swarm), Gbest is replaced by this Pbest.

After the fitness function evaluation phase, the position and velocity of each particle (the amount of location change after an iteration) are updated according to Gbest and Pbest.

The position P_i^{t+1} of particle i at iteration $t+1$ is calculated as:

$$P_i^{t+1} = P_i^t + V_i^{t+1}, \tag{3.1}$$

where P_i^t is the previous position and V_i^{t+1} is the velocity at the current iteration that is determined by

$$V_i^{t+1} = \underbrace{\omega V_i^t}_{\text{inertia}} + \underbrace{c_1 r_1 \left(Pbest_i^t - P_i^t\right)}_{\text{cognitive}} + \underbrace{c_2 r_2 \left(G_{best}^t - P_i^t\right)}_{\text{social}}, \tag{3.2}$$

in which ω is the inertia weight, c_1 and c_2 are the acceleration coefficients, r_1 and r_2 stand for random numbers between 0 and 1. The values of ω, c_1, and c_2 are chosen based on experience. Equation (3.2) delineates the influential factors governing the modification of the particle's velocity: the property of keeping the velocity unchanged (inertia term), the previous particle's knowledge (cognitive term), and the influence from the social (social term).

Following the particles' position and velocity update based on Eqs. (3.1) and (3.2), the stopping criteria are evaluated. If it is satisfied, the algorithm will stop. If not, the algorithm returns to the fitness calculation phase with the new position of each particle in the swarm.

PSO is a theoretically well-founded algorithm for this study due to its many advantages. First, it uses real-valued representations, avoiding the need for complex binary conversions common in other heuristic algorithms such as the GA. Second, its simple arithmetic operations make it easy to understand and implement. Third, its swarm behavior is particularly well-suited for large search spaces, enabling exploration at different scales and communication between particles that share information about their best solutions [1].

Despite numerous advantages, the PSO algorithm also has some drawbacks. One notable drawback is its vulnerability to early convergence, especially when dealing with complex optimization problems with multiple extremes. This phenomenon occurs due to two main inextricably intertwined factors: the inherent characteristics of the functions being optimized and the rapid loss of particle diversity, which can weaken the global search capability of the algorithm. To improve the performance of the PSO algorithm, several innovative PSO variants have been proposed. These variants aim to enhance the capability of PSO to explore the search area and evade local optima, especially for specific problem instances. Generally, approaches to improving the traditional PSO algorithm can be categorized into two primary strategies: (i) hybridizing two existing algorithms, and (ii) introducing a new strategy of tuning the parameters inside the algorithm.

In pursuit of the first strategy, the authors in [2] proposed the gravitational particle swarm optimization algorithm (GPSOA), which hybridizes the PSO with the gravitational search algorithm (GSA). In GPSOA, particle behavior is influenced not only by kinematic parameters such as ω, c_1, c_2, but also by gravitational interactions among them. Another method is the quantum-behaved PSO introduced by [3], which combines quantum mechanics with the conventional PSO algorithm. In the quantum PSO algorithm, each particle's position not only relies on the previous position and current velocity but also depends on the potential field of the particle. These two novel approaches have demonstrated better global search capability compared to the original PSO, particularly in addressing the multi-objective optimization problem of designing wind–thermal power systems [2] or choosing a suitable learning rate for deep belief networks [4].

Although hybridization strategies offer several advantages, they come with limitations. One significant issue arises when combining two algorithms: the increase in the number of parameters that need to be tuned. As the parameter count rises, evaluating the contribution of each parameter to the hybrid algorithm's operation becomes challenging, making effective tuning difficult. Consequently, the second strategy, which involves introducing innovative methods to adjust the algorithm's metrics, is preferred. Various techniques can be applied in this strategy, such as modifying the particle's update formula, altering the particle's velocity, and adjusting the utilization of c_1 and c_2 values [5].

In the PSO—local exploitation orienter (PSO-LEO) approach introduced by [6], the value of ω undergoes an exponential decline as depicted by Eq. (3.3). At the start of the iterations, ω shows a significant drop from ω_{max} to ω_{min}, signifying the algorithm's local search behavior.

$$\omega = \omega_{\min} + \left(\omega_{\max} - \omega_{\min}\right) \times \exp\left(-\alpha \frac{t}{T}\right). \tag{3.3}$$

In the PSO—global exploration orienter (PSO-GEO) approach introduced by [6], where the inertia weight follows Eq. (3.4), ω starts near ω_{max} in the initial iterations and significantly decreases to ω_{min} in later iterations. This employment encourages the algorithm to focus on global exploration early on, thereby improving its ability to locate the global optimum.

$$\omega = \omega_{\max} \times \exp\left(-\left(\frac{t - T\left(1 - \sqrt[\beta]{\ln \frac{\omega_{\max}}{\omega_{\min}}}\right)}{T}\right)^{\beta}\right). \tag{3.4}$$

3.2 Firefly Algorithm

The firefly algorithm (FA) is one of the typical algorithms based on swarm intelligence. It draws inspiration from the natural interaction of fireflies through their illumination. FA is designed to solve global optimization problems, where each firefly individual in the population interacts with others based on their brightness. During the mate selection process, each individual is attracted to and moves toward brighter individuals.

The algorithm is built upon the following three ideal assumptions [7–9]:

- Fireflies are attracted to each other regardless of gender because they belong to a unisex species.
- Attraction is proportional to their brightness, with dimmer fireflies being attracted to brighter ones. However, attraction decreases as the distance between two fireflies increases. Additionally, the brightest firefly's movement is random.
- The brightness of a firefly is dependent upon the objective function's value.

3.2.1 Algorithm Formulation

To properly develop an FA, it is necessary to identify two crucial issues: the variation of light intensity and the attractiveness formula. These problems allow researchers to adjust various FAs in the most suitable manner for the requirements of the problems. In the FA's standardized version, the light intensity I of a firefly represents the solutions proportional to the value of the fitness function. Since both light intensity and attractiveness affect the movement of fireflies in the FA, we need to define their variations. Fundamentally, the attractiveness of a firefly can be assumed to depend on its brightness, which is related to the objective function.

If a firefly fails to find another brighter firefly, it will move randomly in the search space. Assume a firefly population of 4 individuals. The light intensity of the ith individual is denoted as I_i. Suppose $I_1 > I_2$ and $I_3 > I_4$. Individuals 2 and 4 move closer to the brighter individuals, while individuals 1 and 3 move randomly, as illustrated in Figure 3.1.

Because the light intensity of the ith individual at a certain distance r_i from the light source follows the inverse square law, the attractiveness of a firefly decreases with its distance from another firefly. This means that the light intensity I_i decreases as the distance r_i increases, in accordance with $I_i \propto \frac{1}{r_i^2}$. Additionally, the atmosphere to absorb light becomes weaker with the increasing distance. These combined factors limit the visibility of most fireflies to a restricted distance, which implies that effective communication only occurs among individuals in proximity, leading to strong interactions among them.

The brightness of a firefly is typically proportional to the objective function of the problem f_{obj}, which can be expressed as [10]

$$I_i \propto f_{obj}(x_i), \tag{3.5}$$

where I_i and x_i denote the brightness and the position corresponding to the ith firefly of the population.

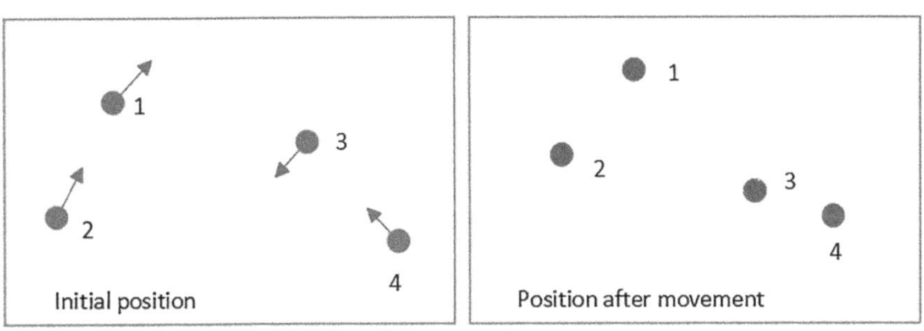

Figure 3.1 Example of the movement of a firefly population.

The light intensity I_i follows the inverse square law, which can be given by [11]

$$I_i = \frac{I_S}{r_i^2}, \tag{3.6}$$

where I_S is the light intensity at the source and r is the distance from the considered point to the light source.

In an environment where the light absorption coefficient γ is constant, the light intensity varies with distance r_i according to

$$I_i = I_0 e^{-\gamma r_i}. \tag{3.7}$$

In light of (3.6), to avoid the undefined singularity at $r_i = 0$, the light intensity can be approximated in the form of a Gaussian as [10]

$$I_i = I_0 e^{-\gamma r_i^2}. \tag{3.8}$$

Additionally, the attractiveness of a firefly is perceived by other fireflies. A firefly with lower brightness will be attracted to the one with higher brightness, and each firefly at different positions will have its own unique attractiveness value β. Since this attractiveness also varies depending on distance, β strongly resembles I. While intensity I is considered an absolute measure of light emitted by a firefly, attractiveness β measures the relative light to be seen and evaluated by other fireflies. Therefore, the attractiveness of a firefly, which is proportional to the light intensity emitted by it, can be defined as [8, 12]

$$\beta = \beta_0 e^{-\gamma r_{ij}^2}. \tag{3.9}$$

In (3.9), the distance r_{ij} between the ith firefly at position X_i and the jth firefly at position X_j is calculated as [9]

$$r_{ij} = ||X_i - X_j|| = \sqrt{\sum_{k=1}^{d} (x_k^i - x_k^j)^2}, \tag{3.10}$$

where ||.|| is the symbol for distance in Descartes coordinate, d is the dimension of the search space, x_k^i and x_k^j are the coordinates in the kth dimension of the ith and jth firefly, respectively.

The movement of the ith individual when attracted by a brighter jth individual is defined as [8, 9]

$$X_i^{t+1} = X_i^t + \beta_0 e^{-\gamma r_{ij}^2}(X_j^t - X_i^t) + \alpha \epsilon_i^t, \tag{3.11}$$

where X_i^{t+1} is the updated position of the ith firefly, X_i^t and X_j^t are the initial position of the ith and jth firefies, α represents the coefficient affecting the random motion of fireflies, and ϵ_i^t represents a random number following a normal distribution. The movement of a firefly in (3.11) consists of three consecutive parts, namely the current position X_i^t of the ith firefly, the attraction of another jth firefly that is more attractive, and the random movement with the coefficient ϵ_i^t.

The algorithm compares the attractiveness of the firefly's new position with the old one. The fireflies' positions can be sequentially updated by comparing and updating each pair of fireflies after each iteration. If the new position generated is more attractive, the firefly will move to the new position; otherwise, it will remain at its current position.

Local search algorithms often face the risk of getting trapped at a local optimum while the global optimum lies outside the search scope. To address this issue, randomness is often introduced into an algorithm to allow it to jump out of such positions. Random components can take various forms, such as simple randomization by sampling randomly within the search space or by taking random jumps.

The movement expression of a firefly in (3.11) reflects a random step biased toward brighter firefly individuals. In the case where the firefly individual has the highest brightness intensity ($\beta_0 = 0$), the expression simplifies to a random movement step as

$$X_i^{t+1} = X_i^{t+1} + \alpha\epsilon_i^t. \tag{3.12}$$

In FA, fireflies interact with each other through their attractiveness, allowing them to explore the search space via the objective function. This method is more effective than a random search following a normal distribution because each individual in the swarm explores their space based on the results obtained by other individuals, while still applying their own random movement steps. The combination of individual movement and position updates based on brightness helps the algorithm balance exploration and exploitation.

According to the theory of FA, a firefly tends to move toward other more attractive individuals. Therefore, this algorithm has a noticeable advantage, namely automatic congregation. Indeed, fireflies may divide the population and automatically congregate into smaller groups because the attraction of nearby individuals is stronger than the attraction of those at a distance; thus, one can expect each group of individuals to concentrate on promising areas of the search space. This advantage makes it particularly suitable for multi-objective global optimization problems, including analog sizing optimization. The reason is that focusing on or exploiting potential regions after exploring the vast analog design space might bring about better solution quality, faster convergence, and reduced optimization time and resources.

3.2.2 Implementation

The implementation of FA is described in Algorithm 3.2 and subsequent clarification for each step.

Algorithm 3.2: Firefly Algorithm (FA)

Initialize an m-individual population $P_{FA} = \{X^1, X^2, \dots, X^i, \dots, X^m\}$;
Calculate the fitness values $f_{obj}(X^i)$ of each firefly;
while *(!stop_condition)* **do**
  for $p = 1 : m$ **do**
    for $q = 1 : m$ **do**
      if $f_{obj}(X^p)$ *is more optimal than* $f_{obj}(X^q)$ **then**
        Move X^q closer to X^p based on (3.11);
      Move the best firefly X^* randomly based on (3.12);
      Calculate the new fitness values of each firefly;
return X^*;

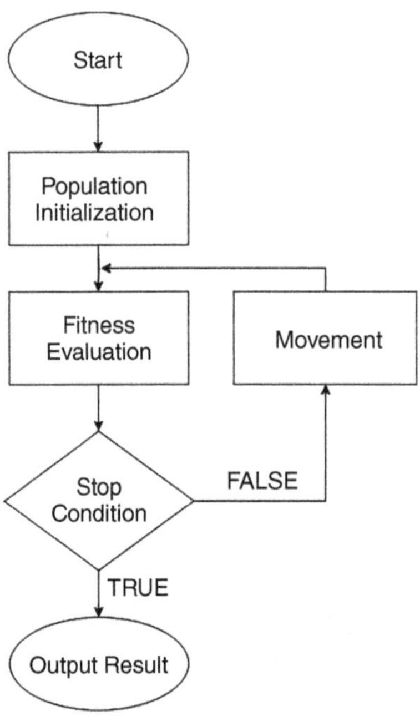

Figure 3.2 FA's flowchart.

The FA's implementation steps can be illustrated by the flowchart in Figure 3.2.

3.3 Cuckoo Search

Introduced in 2009 by Xin-She Yang and Suash Deb, the cuckoo search (CS) algorithm has been recognized as a pioneering metaheuristic optimization technique that outperforms earlier algorithms such as PSO and GA. The CS algorithm is inspired by the brood parasitism behavior observed in cuckoo species, which involves cooperation and direct conflict between host birds and intruding cuckoos, where host birds may remove alien eggs or abandon their nest to establish a new one.

The CS algorithm leverages this natural strategy to improve the search for the optimal global solution. In the implementation of the CS algorithm for optimization tasks, the algorithm starts by generating n candidate solutions, conceptualized as nests, each with coordinates representing potential solutions. Each nest is evaluated through the fitness function, which determines its suitability by comparing the fitness at a given coordinate x_0 with the value of the objective function when the design variable x is equal to x_0.

The proposal of the CS algorithm draws directly from the aforementioned brood parasitism behaviors. To streamline the algorithm's construction and application, it incorporates three idealized assumptions:

1) One egg is laid at a time, and the host nest is selected randomly.
2) The nest with the most favorable conditions for hatching is retained for the subsequent generation.
3) There exists a fixed number of nests, and the probability that a parasitic egg is detected is $p_a \in [0, 1]$. If an alien egg is discovered, the current nest will be replaced by a new nest in a new position.

The basic steps of the CS algorithm can be summarized in Figure 3.3.

CS employs a blend of local random walk and global explorative random walk. A global random walk relying on the Lévy flight to determine the new position of the host nest i in the $t+1$ iteration can be expressed as the Mantegna algorithm as

$$x_i^{t+1} = x_i^t + \alpha(x_i^t - x_{best}) \oplus L(\lambda), \tag{3.13}$$

where x_i^t is the position of the ith host nest at the tth iteration, x_{best} is the current optimal solution, and α is the step-length control factor. $L(\lambda) = \mu/|v|^{1/\beta}$ is the Lévy flight with $\beta = 1.5, \mu \sim N(0, \sigma_u^2), v \sim N(0, 1), \sigma_u = \frac{\Gamma(1+\beta) * \sin(\pi\beta/2)}{\Gamma[(1+\beta)/2] * \beta * 2^{(\beta-1)/2}}$. The product $\oplus$ means entry-wise multiplication.

Following the global random walk, for each nest, a random number r is generated. If $r > p_a$, the cuckoo's parasitic egg is considered to have been detected, and a local random walk is then used to replace the old solution with a new one.

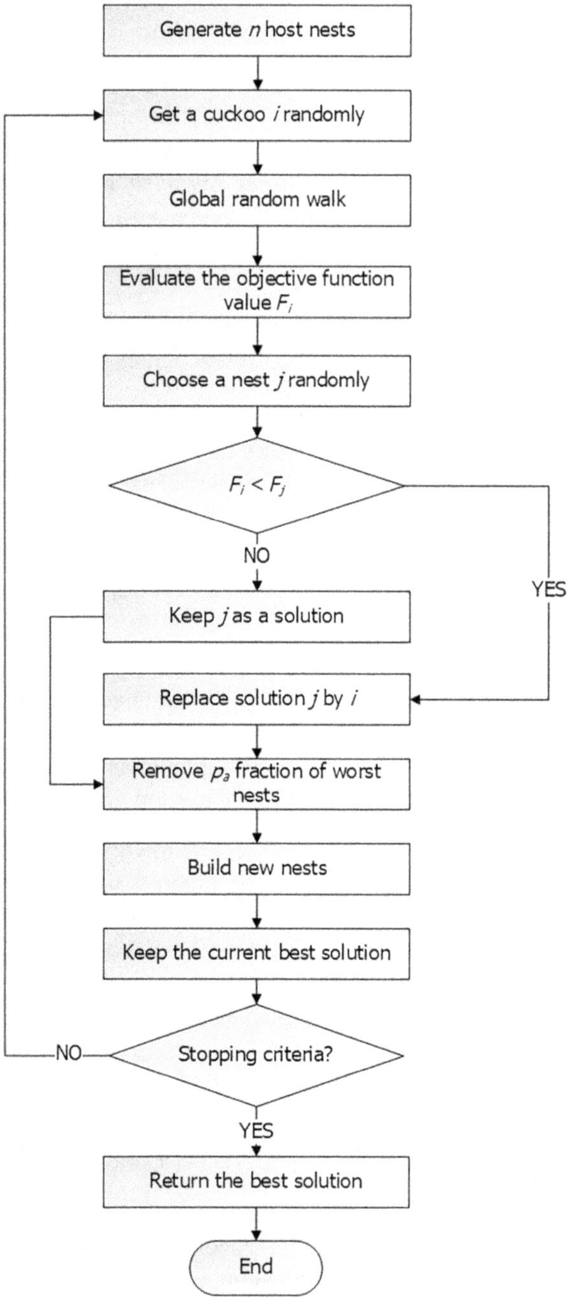

Figure 3.3 The flowchart of CS algorithm.

If $r < p_a$, the parasitic egg is not considered discovered, and the host nest will not be abandoned. This process can be summarized as

$$x_i^{t+1} = \begin{cases} x_i^t + \alpha s \otimes H(p_a - r) \otimes (x_j^t - x_k^t), & r > p_a \\ x_i^t, & r \leq p_a, \end{cases} \tag{3.14}$$

where x_j^t and x_k^t are two different solutions selected randomly by random permutation, H is the Heaviside function, r is a random number drawn from a uniform distribution, s is the step size, and $\otimes$ is the entry-wise product of two vectors.

Compared with Brownian random walks, Lévy flights are more effective in exploring unknown, expansive search areas. This efficiency can be attributed to the fact that the variance of Lévy flights, $\sigma^2(t) \sim t^{3-\beta}$ with $1 \leq \beta \leq 2$, increases at a much faster rate than the linear relationship (i.e., $\sigma^2(t) \sim t$) of Brownian random walks.

3.4 Bat Algorithm

Bat algorithm (BA), constructed by Xin-She Yang in 2010 as a meta-heuristic optimization algorithm [12, 13], is mainly inspired by the widely recognized phenomenon called echolocation of microbats [11]. It is the first algorithm of the animal-behavior-inspired group to use frequency tuning together with varying pulse emission rate and loudness to diversify the solutions within the population [14].

The echolocation mechanism is utilized by microbats to spot prey, avoid obstacles, and navigate their habitat in darkness. By estimating the time lag between emitting and receiving echoes, the time difference between their two ears and the alterations in echo loudness, microbats construct a three-dimensional perception of the surroundings, including the target's distance and orientation, the type of prey and even its speed [12, 15]. Furthermore, the sound pulse can be adjusted according to the microbats' hunting strategy for each type of prey. This usually means that as the microbat approaches its prey, its pulse rate increases while its loudness decreases and vice versa [16].

In developing BA, several idealized rules of the above-mentioned echolocation characteristics of microbats are employed as [12, 15]

- Echolocation allows all bats to sense distance and magically differentiate between food and background objects.
- At a position x_i, the bat exhibits a random flight behavior with velocity v_i. They can self-modulate the frequency (or wavelength) of their emitted signals of their emitted pulses and pulse emission rate $r \in [0, 1]$, based on the proximity of their target.

- Loudness, despite varying in different ways, can be assumed to span from a minimum value A_{min} to a large positive value A_0.

BA, a population-based optimization algorithm, utilizes an iterative process to update each individual in the population toward the global best solution and their corresponding personal best solution [14]. This aspect can be suitable for analog sizing optimization since it provides an oriented search direction in the complex and possibly multi-dimensional design space, which might result in better efficiency of the overall analog sizing optimization procedure.

For a population of n microbats, at each iteration t, the D-dimensional position vector x_i^t and velocity v_i^t follow an update mechanism given by

$$f_i = f_{min} + (f_{max} - f_{min})\beta, \tag{3.15}$$

$$v_i^{t+1} = v_i^t + (x_i^t - x_*)f_i, \tag{3.16}$$

$$x_i^{t+1} = x_i^t + v_i^{t+1}, \tag{3.17}$$

where x_* is the currently optimal solution, $\beta \in [0, 1]$ is a uniformly distributed random vector, and $f_i \in [f_{min}, f_{max}]$ is the sonic frequency of the bat.

In terms of the local search, after selecting a solution from the currently best individuals, a local random walk is employed to generate a bat individual, which is expressed as

$$x_i^{t+1} = x_i^t + x_* f_i, \tag{3.18}$$

where $x_* f_i$ is the local solution based on the currently best solution.

The loudness A_i and pulse emission rate r_i are adjusted according to (3.19) and (3.20) as

$$A_i^{t+1} = \alpha A_i^t, \tag{3.19}$$

$$r_i^{t+1} = r_i^0(1 - e^{-\gamma t}), \tag{3.20}$$

where $\alpha \in [0, 1]$ is the acoustic loudness attenuation coefficient, $\gamma > 0$ is the pulse frequency enhancement coefficient, and r_i^0 is the initial pulse frequency of microbat i.

From (3.19) and (3.20), for any $0 < \alpha < 1$ and $\gamma > 0$, we have

$$A_i^t \longrightarrow 0, r_i^t \longrightarrow r_i^0 \quad \text{as} \quad t \longrightarrow \infty. \tag{3.21}$$

Eq. (3.21) indicates that once the microbat finds its prey, it will gradually reduce the loudness of its pulse emission while increasing its pulse emission rate. This is consistent with the above-mentioned characteristics of the microbat's hunting strategy.

BA's essential procedure, based on the demonstrated idealized rules, is captured in the pseudocode of Algorithm 3.3.

Algorithm 3.3: Bat Algorithm (BA)

Initialize the population of microbats using x_i and v_i ($i = 1, 2, \ldots, n$);
Initialize f_i, r_i, A_i;
Evaluate the values of the objective function f_{obj} for each individual;
while *!stop_condition* **do**
  Adjust frequency, velocity, position according to (3.15), (3.16), (3.17);
  Generate *rand*, a uniformly random value in the range [0, 1];
  if $rand > r_i$ **then**
    Generate a local solution based on the currently best solution for the selected best solution;
  Generate a new solution by a local random walk as in (3.18);
  if $rand > A_i$ *or* $f_{obj}(x_i) < f_{obj}(x_*)$ **then**
    Accept the new solutions;
    Increase r_i and reduce A_i;
  Sort the microbats based on the values of their objective function and find x_*;
return x_*;

3.5 Flower Pollination Algorithm

Flower pollination algorithm (FPA), also called flower algorithm, is a metaheuristic and bionic optimization algorithm reflecting the natural pollination of flowering plants [17]. In nature, it is estimated that 80% of the plant species are flowering species, which have been ubiquitous in the landscape since the Cretaceous period [12]. To preserve this dominance among plants, flowering plants need to maintain their primary aim of reproduction, which is achieved through pollination. In fact, during pollination, pollen or the male gamete is carried to the stigma to fertilize the female gamete [18].

This pollen transfer is often associated with pollinators such as insects, which have established a unique flower-pollinator partnership called flower constancy [19]. The phenomenon involves the selective visit to certain flower species while disregarding others thanks to the flowers' charming petals, fragrance, and nectar. Consequently, enhanced pollen transfer to conspecific or identical plants maximizes the reproduction of the same flower species, which is referred to as flower constancy's evolutionary advantage [12]. Moreover, flower constancy can also do wonders for the pollinators thanks to the guarantee of an available nectar supply as well as the reduced effort for learning, switching, or discovering new yet unpredictable flower species.

Pollination can be categorized into two major types, namely abiotic and biotic pollination with their respective proportion of 10% and 90% in the world of flowering plants [20, 21]. Abiotic pollination, in which grass makes a good example, happens infrequently as pollen is transferred by gravity, wind, or diffusion in water rather than by pollinators with reduced accuracy [19]. On the other hand, biotic pollination occurs when pollinators such as animals or insects carry pollen from one flower to another, which might lead to the flower's fertilization. Furthermore, there exists another classification including cross-pollination and self-pollination. The former type involves the transfer of pollen from one flower to another of a different plant, while the latter indicates the transfer of pollen from one flower to another of the same plant, which often happens in the case of no reliable pollinators.

From the optimization viewpoint, biotic and cross-pollination can take place over a long distance thanks to the long flying distance by pollinators such as bees, flies, birds, and bats; hence, they can be regarded as global pollination. In this global pollination process, flight patterns of bees and birds may involve Lévy flight behavior, with corresponding Lévy-distributed jumps or flight distance steps. By contrast, self-pollination and abiotic pollination can be identified as forms of local pollination due to their short travel distance of pollen. Additionally, flower constancy can be viewed as an incremental step based on the similarity or difference between two flowers. Last but not least, the ultimate aim of flower pollination, specifically natural selection and the most effective reproduction of plants, makes an analogy with the objective of an optimization problem. All these interactive processes and factors for the optimal reproduction of flowering plants are the primary motivation for the formulation of the FPA.

In order to mathematically model and describe FPA, the following idealized assumptions are utilized [17, 22]:

- Biotic and cross-pollination can be regarded as global pollination with pollinators transferring pollen according to Lévy flight. This phase corresponds to the global search or exploration stage of candidate solutions in the solution search space with regard to Lévy flight.
- Local pollination, implemented by abiotic and self-pollination, can be considered the local search or exploitation stage toward better solutions in the solution search space.
- Flower constancy, governed by the reproduction probability rule replicating pollinators' tendencies to fertilize certain flowers, can be taken into account owing to the similarity of two flowers.
- The switching probability $p \in [0, 1]$ controls the switching process between global and local pollination since flowers in nature may not occur far away from each other and are more likely to be pollinated by local gametes.

In FPA, firstly, a population of n individuals with d dimensions is initialized. Subsequently, the fitness values or the values of the objective function f_{obj} of all individuals are evaluated to find the currently optimal solution. To generate new solutions, the switch probability p needs to be taken into account to determine between global and local pollination. An elitist selection is followed by this pollination step to compare the new and past solutions. FPA's implementation is illustrated by the pseudocode of Algorithm 3.4 while its three major processes, namely global pollination, local pollination, and elitist selection, are further demonstrated in Sections 3.5.1–3.5.3.

Algorithm 3.4: Flower Pollination Algorithm (FPA)

Initialize the switching probability $p \in [0, 1]$;
Generate a random population of flowers;
Evaluate the values of the objective function f_{obj} of each flower and determine the global best solution x_*;
$t \leftarrow 1$;
while $t < T$ **do**
    Generate *rand*, a uniformly random value in the range $[0, 1]$;
    if $rand < p$ **then**
        Perform global pollination according to (3.22);
    else
        Perform local pollination according to (3.26);
    Evaluate the values of the objective function f_{obj} of new solutions;
    Perform elitist selection according to (3.27);
return x_*

3.5.1 Global Pollination

In the global pollination process, it is assumed that each plant has a single flower, and each flower produces one pollen gamete [23]. As a result, a candidate solution x_i plays the role of a single flower [22]. In global pollination, pollens are carried through long distances by pollinators with Lévy flight characteristics; hence, the first rule can be expressed mathematically as

$$x_i^{t+1} = x_i^t + \gamma L(x_i^t - x_*), \tag{3.22}$$

where x_i^t, x_i^{t+1} represents the ith pollen at iteration $t, t+1$, x_* is the global best solution, and γ stands for the scaling factor. L indicates the step size drawn from

Lévy flight and is given by

$$L \sim \frac{\lambda\Gamma(\lambda)\sin(\frac{\pi\lambda}{2})}{\pi} \times \frac{a}{s^{1+\lambda}}, \quad (|s| \to \infty, s \gg s_0 > 0), \tag{3.23}$$

where $\Gamma(\lambda)$ is the standard gamma function with an index λ, a is a control parameter for the tail amplitude of the distribution (a usually equals to 1), s and s_0 represent the respective Lévy step size and the minimum step size, respectively.

The Lévy step size s in (3.23), which is usually large, can be written as

$$s = \frac{U}{|V|^{\lambda-1}}, \tag{3.24}$$

where U and V are two random samples drawn from a Gaussian normal distribution with zero mean, and standard deviations σ_u and σ_v given by

$$U \sim (0, \sigma_u^2), V \sim (0, \sigma_v^2),$$
$$\sigma_u = \left[\frac{\Gamma(1+\lambda)}{\lambda\Gamma((1+\lambda)/2)} \times \frac{\sin(\frac{\pi\lambda}{2})}{2^{(\lambda-1)/2}}\right]^{1/\lambda}, \tag{3.25}$$

where the standard deviation σ_v is set to 1 as σ_u and σ_v cannot be chosen independently for any value of.

3.5.2 Local Pollination

Local pollination, the transfer of pollen via abiotic means such as wind, occurs within a relatively short distance. Consequently, the neighboring exchange of pollen and flower constancy, which can also be considered the local search or exploitation phase, is enabled. Similar to global pollination, local pollination described in the second rule and flower constancy described in the third rule can be formulated as

$$x_i^{t+1} = x_i^t + \epsilon(x_j^t - x_k^t) \quad (j \neq i, k \neq i, j \neq k), \tag{3.26}$$

where x_j^t and x_k^t are two randomly selected jth and kth solution, and ϵ is a uniformly distributed random number in the range $[0, 1]$.

3.5.3 Elitist Selection

The newly generated solution x_i^{t+1} from the global or local pollination process is compared with its previous solution x_i^t. If the fitness value of x_i^{t+1} is more optimal than x_i^t, the past solution is then replaced; otherwise, x_i^t remains in the population. This procedure is called elitist selection, which is expressed as

$$x_i^{t+1} = \begin{cases} x_i^{t+1}, & \text{if } f_{obj}(x_i^{t+1}) \text{ is more optimal than } f_{obj}(x_i^t), \\ x_i^t, & \text{otherwise.} \end{cases} \tag{3.27}$$

3.6 Ant Colony Optimization

Ant colony optimization (ACO), developed by Marco Dorigo in the early 1990s, is a meta-heuristic algorithm to approximately solve complex combinatorial optimization problems within a practical time frame [24]. It takes inspiration from the natural foraging habits of ant colonies, involving finding the shortest possible path in the shortest amount of time to a food source [25]. In constructing the ACO's optimization model, the indirect communication using pheromone, a chemical substance, among ants plays a crucial role as a means to reflect ants' search experience [26].

The ants' process of searching for food can be described as follows. In the beginning, the ants randomly select tours to explore the area surrounding their nest. Until a food source is discovered, they assess the quantity and quality of the food and then carry some of it back to the nest. On the way back to the nest, the ant deposits some pheromone with an evaporation feature to mark the path. The amount of laid pheromone on the edges of the tours, which may be influenced by the distance, quantity, and quality of the food source, helps orientate other ants to the food source. Subsequently, the ants are more likely to follow paths with abundant pheromones, leading to even more pheromones being deposited on those routes. As a result, the amount of pheromones will be higher for the best ants that find better food sources, attracting more ants to follow their trails. In the end, as shown in [27], this positive feedback mechanism enables all the ants to follow the shortest path between their nest and the food source [24, 26]. Specifically in analog sizing optimization, the mentioned shortest path as a result of the positive feedback mechanism might enable an efficient search for circuit components' dimensions in the broad and multi-dimensional design space. During the search process using the positive feedback mechanism, there is a trade-off between exploration and exploitation, which is represented by the random search for the food source and the share of information about the food source via pheromone, respectively [25].

At the beginning of the ACO algorithm, a grid search space is established, in which n nodes are chosen. Each ant is randomly positioned at a starting node. Moreover, all ants are initialized with a constant amount of pheromone and then will search randomly for a food source and return to their nest. This gives them a perception of their search space and can therefore share the information with other ants. At each node, a transition rule is employed by each ant to move to an adjacent node and construct a complete path. The probability of choosing the next node is expressed as

$$P_{ij}^{k} = \begin{cases} \dfrac{[\tau_{ij}]^{\alpha}[\eta_{ij}]^{\beta}}{\sum_{l \in N_i^k} [\tau_{il}]^{\alpha}[\eta_{il}]^{\beta}} & \text{if } j \in N_i^k, \\ 0 & \text{otherwise,} \end{cases} \tag{3.28}$$

where P_{ij}^k illustrates the state transition probability from node i to node j of the kth ant, τ_{ij} is the amount of pheromone associated with the edge (i,j) representing exploration, η_{ij} is the heuristic function related to visibility (distance) representing exploitation, and N_i^k denotes the unvisited food sources for the kth ant.

In (3.28), α and β are positive real parameters that determine the relative importance of pheromone compared to heuristic information and regulate the influence of the pheromone value $[\tau_{ij}]^\alpha$ and the heuristic value $[\eta_{ij}]^\beta$, respectively. Since it is clear from (3.28) that a larger value of either α or β corresponds to a greater contribution of either the pheromone strength or visibility (distance) to the transition probability P_{ij}^k. Hence, α is called the pheromone exponent, while β is called the heuristic exponent.

The heuristic function η_{ij} is defined as the inverse of the distance between nodes i and j as

$$\eta_{ij} = \frac{1}{d_{ij}}, \tag{3.29}$$

where d_{ij} is the Euclidean distance between nodes i and j.

According to the ant's foraging behavior, the pheromone is deposited along their path to attract other ants to follow, and this pheromone evaporates. The process of deposition and evaporation is referred to as the local pheromone update, which is given by

$$\tau_{ij} = (1-\rho)\tau_{ij} + \rho\tau_0, \tag{3.30}$$

where τ_0 stands for the initial pheromone value and $\rho \in [0;1]$ represents the evaporation rate.

In the final stage, the ACO algorithm evaluates the solutions of all the ants and records the best one. The ant with the best solution then deposits a pheromone along its path (global update) to guide other ants in future iterations, as expressed as

$$\tau_{ij} = (1-\rho)\tau_{ij} + \rho\Delta\tau_{ij}, \tag{3.31}$$

in which

$$\Delta\tau_{ij} = \begin{cases} \frac{1}{L_{gb}} & \text{if edge } (i,j) \text{ is the best path,} \\ \tau_{ij} & \text{otherwise,} \end{cases} \tag{3.32}$$

3.7 Case Study: PSO and Cuckoo Search Algorithm Implementation in Band-gap Reference Circuit Design[1]

This section explores the method of employing various renowned metaheuristic optimization algorithms aimed at maximizing the power supply rejection ratio (PSRR) parameter of the BGR circuit. The structure of a BGR circuit, as proposed by [6], is illustrated in Figure 3.4. The sizing framework is illustrated in Figure 3.5, comprising two main components: the circuit block and the algorithm block. The circuit block includes the schematic representation of the BGR circuit, from which the netlist file is generated and subsequently imported into the Spectre simulator. In the algorithm block, the metaheuristic algorithms are implemented using the Python programming language. In this case study, five metaheuristic algorithms are utilized, which encompass the conventional PSO, the PSO-GEO introduced in [6], the CS algorithm, and the FA algorithm introduced in this chapter. Serving as a communication channel between the circuit block and the algorithm block is

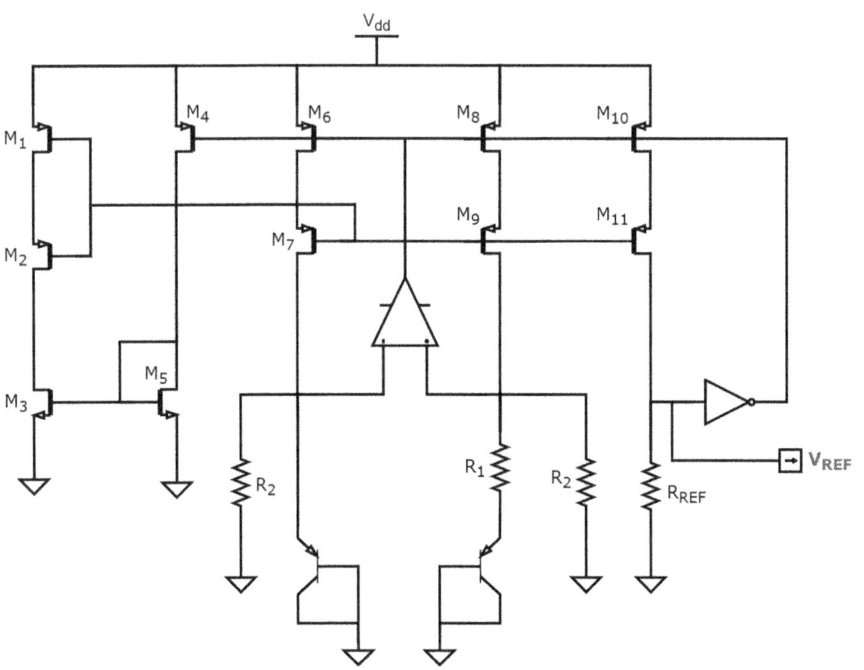

Figure 3.4 Schematic of BGR circuit.

1 This subsection has been published as part of [6].

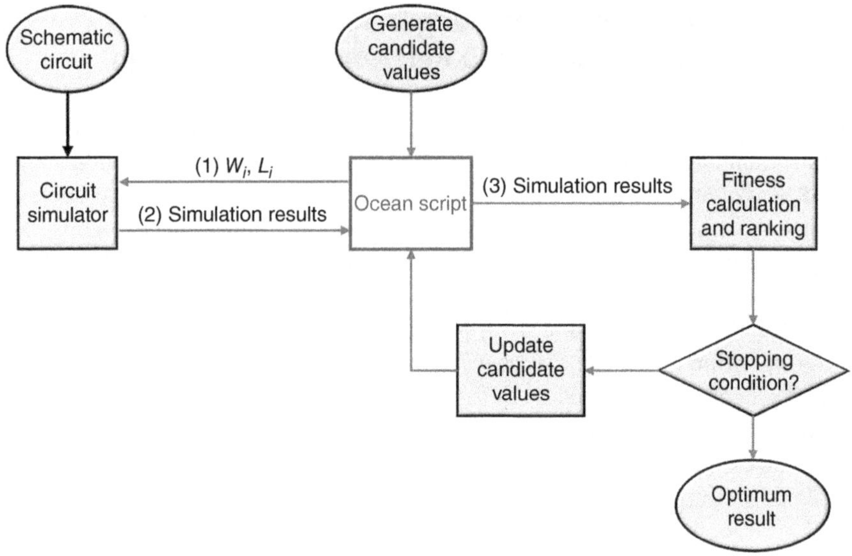

Figure 3.5 Framework for implementing the metaheuristic algorithms in designing BGR circuit.

a script written in the Ocean language. This script is utilized to automate the running of the Spectre simulator, facilitating the extraction of the value of the objective function and the constraint functions. Identical initial positions and control parameters are employed to ensure a fair comparison among the five algorithms. Due to the random nature of metaheuristic algorithms, each method undergoes 10 independent runs, each comprising 100 iterations for statistical robustness.

Table 3.1 presents the optimization results obtained by conducting 10 trials for each of the five metaheuristic algorithms mentioned above. As can be seen, due to the heuristic nature of these algorithms, the fitness values achieved in each trial vary not only between different algorithms but also between multiple trials of the same algorithm. The unique mathematical foundation of each algorithm endows it with specific strengths and weaknesses. The three PSO-based algorithms tend to have lower success rates than the others, probably because the PSO algorithm focuses on intensifying the search around the best fitness value (G^*) identified within the swarm. On the other hand, the CS and FA algorithms, whose random walks are formulated by the Lévy flight and Gaussian distribution, respectively, are more inclined toward exploring the search space rather than exploiting it. This trade-off between exploration and exploitation explains why the CS algorithm boasts a higher success rate than PSO, and why traditional PSO and PSO-GEO achieve solutions with better values and smaller standard deviations of the fitness function, compared to the CS algorithm, as presented in Table 3.1.

Table 3.1 Fitness function obtained from ten executions of five different algorithms.

Execution	Traditional PSO	GPSOA	PSO-GEO	CS	FA
1	−97.592	901.214	−99.064	−86.582	−98.726
2	905.991	−97.595	−99.725	−97.762	−98.620
3	904.555	901.800	903.633	−97.565	−99.081
4	904.600	907.617	−100.234	−97.551	−98.528
5	898.464	903.237	899.591	−97.965	−98.612
6	−99.382	902.246	−99.794	−98.436	−97.865
7	899.296	900.008	−99.670	−98.523	−99.047
8	−100.299	905.598	−97.592	−93.487	−96.372
9	−97.700	−98.772	904.454	−98.752	−98.517
10	−98.000	−99.042	−99.941	−97.882	−98.372
Success rate	5/10	3/10	7/10	10/10	10/10
Minimum fitness value	−100.299	−99.042	−100.234	−98.752	−99.081
Mean value (success cases only)	−98.595	−98.470	−99.431	−96.451	−98.374
Std.dev (success cases only)	1.066	0.628	0.819	3.581	0.742

References

1 A. Pérez-González, O. Begovich-Mendoza, and J. Ruiz-León, "Modeling of a Greenhouse Prototype Using PSO and Differential Evolution Algorithms Based on a Real-Time LabView™Application," *Appl. Soft Comput.*, vol. 62, pp. 86–100, 2018.

2 S. Jiang, C. Zhang, W. Wu, and S. Chen, "Combined Economic and Emission Dispatch Problem of Wind-Thermal Power System Using Gravitational Particle Swarm Optimization Algorithm," *Math. Probl. Eng.*, vol. 2019, no. 1, p. 5679361, 2019.

3 J. Sun, W. Fang, X. Wu, V. Palade, and W. Xu, "Quantum-Behaved Particle Swarm Optimization: Analysis of Individual Particle Behavior and Parameter Selection," *Evol. Comput.*, vol. 20, no. 3, pp. 349–393, 2012.

4 C. Zhang, Y. He, L. Yuan, and S. Xiang, "Analog Circuit Incipient Fault Diagnosis Method Using DBN Based Features Extraction," *IEEE Access*, vol. 6, pp. 23053–23064, 2018.

5 D. Wang, D. Tan, and L. Liu, "Particle Swarm Optimization Algorithm: An Overview," *Soft Comput.*, vol. 22, pp. 387–408, 2018.

6 T. Hoang, T. Q. Nguyen, L. Zhang, and T. Q. Duong, "Novel Methods for Improved Particle Swarm Optimization in Designing the Bandgap Reference Circuit," *IEEE Access*, vol. 11, pp. 139964–139978, Dec. 2023.

7 F. Zitouni, S. Harous, and R. Maamri, "A Novel Quantum Firefly Algorithm for Global Optimization," *Arabian J. Sci. Eng.*, vol. 46, pp. 86–100, Apr. 2021.

8 I. Fister, I. Fister, X.-S. Yang, and J. Brest, "A Comprehensive Review of Firefly Algorithms," *Swarm Evol. Comput.*, vol. 13, pp. 34–46, Dec. 2013.

9 V. Kumar and D. Kumar, "A Systematic Review on Firefly Algorithm: Past, Present, and Future," *Arch. Comput. Methods Eng.*, vol. 28, pp. 3269–3291, Sep. 2020.

10 X.-S. Yang, *Nature Inspired Metaheuristic Algorithms*. Bristol, England, United Kingdom: Luniver Press, 2010.

11 F. Rothlauf, *Optimization Methods*. Berlin, Heidelberg: Springer Berlin Heidelberg, 2011, pp. 45–102.

12 X.-S. Yang, *Optimization Techniques and Applications with Examples*. New Jersey: Wiley, 2018.

13 X.-S. Yang, *A New Metaheuristic Bat-Inspired Algorithm*. Berlin, Heidelberg: Springer Berlin Heidelberg, 2010, pp. 65–74.

14 J. Senthilnath, S. Kulkarni, J. A. Benediktsson, and X.-S. Yang, "A Novel Approach for Multispectral Satellite Image Classification Based on the Bat Algorithm," *IEEE Geosci. Remote Sens. Lett.*, vol. 13, no. 4, pp. 599–603, Mar. 2016.

15 L. F. Zhu, J. S. Wang, H. Y. Wang, S. S. Guo, M. W. Guo, and W. Xie, "Data Clustering Method Based on Improved Bat Algorithm with Six Convergence Factors and Local Search Operators," *IEEE Access*, vol. 8, pp. 80536–80560, Apr. 2020.

16 D. Singh, R. Salgotra, and U. Singh, "A novel modified bat algorithm for global optimization," in *2017 International Conference on Innovations in Information, Embedded and Communication Systems (ICIIECS)*, Coimbatore, India, Mar. 2017, pp. 1–5.

17 M. Lei, Y. Zhou, and Q. Luo, "Enhanced Metaheuristic Optimization: Wind-Driven Flower Pollination Algorithm," *IEEE Access*, vol. 7, pp. 111439–111465, Aug. 2019.

18 R. Frankel, E. Galun, and E. Galun, *Pollination Mechanisms, Reproduction, and Plant Breeding*, Monographs on Theoretical and Applied Genetics. Berlin: Springer-Verlag, 1977.

19 M. Abdel-Basset and L. A. Shawky, "Flower Pollination Algorithm: A Comprehensive Review," *Artif. Intell. Rev.*, vol. 52, no. 4, pp. 2533–2557, Dec. 2019.

20 Y. Xiao, Y. Wu, and F. Yang, "Robust Visual Tracking Based on a Modified Flower Pollination Algorithm," *IEEE Access*, vol. 9, pp. 157458–157467, Nov. 2021.

21 E. Nabil, “A Modified Flower Pollination Algorithm for Global Optimization,” *Expert Syst. Appl.*, vol. 57, pp. 192–203, Sep. 2016.

22 S. Lalljith, I. Fleming, U. Pillay, K. Naicker, Z. J. Naidoo, and A. K. Saha, “Applications of Flower Pollination Algorithm in Electrical Power Systems: A Review,” *IEEE Access*, vol. 10, pp. 8924–8947, Jan. 2022.

23 X.-S. Yang, “Flower pollination algorithm for global optimization,” in *Proc. Unconventional Computation and Natural Computation*, Berlin, Heidelberg, Germany, Dec. 2012, pp. 240–249.

24 M. Dorigo and C. Blum, “Ant Colony Optimization Theory: A Survey,” *Theor. Comput. Sci.*, vol. 344, no. 2, pp. 243–278, Nov. 2005.

25 F. Dahan, K. E. Hindi, A. Ghoneim, and H. Alsalman, “An Enhanced ant Colony Optimization Based Algorithm to Solve QoS-Aware Web Service Composition,” *IEEE Access*, vol. 9, pp. 34098–34111, Feb. 2021.

26 J. He, X. Sun, W. Li, and J. Che, “A New Pheromone Update Strategy for Ant Colony Optimization,” *J. Intell. Fuzzy Syst.*, vol. 32, pp. 1–10, Mar. 2017.

27 J.-L. Deneubourg, S. Aron, S. Goss, and J. Pasteels, “The Self-Organizing Exploratory Pattern of the Argentine Ant,” *J. Insect Behav.*, vol. 3, p. 159, Mar. 1990.

4

Human-Behavior-Inspired Algorithms in Analog Circuit Sizing Optimization

Chapters 2 and 3 have provided insights into evolutionary and animal-behavior-inspired algorithms for analog sizing optimization. Both approaches seek improvement in their intended objectives, either evolutionary processes or swarm activities, which cater for the purpose of the optimization discipline. In a similar manner, human behavior, which refers to cognitive processes employed to accomplish a specific task, can also be a source of inspiration for optimization algorithms. Indeed, a variety of algorithms have been proposed and widely applied based on the imitation of human-problem-solving strategies. This chapter introduces Ali Baba and the forty thieves (AFT) algorithm, drawer algorithm (DA), political optimizer (PO), and war strategy optimization (WSO), which are considered some of the most suitable candidates for addressing the analog sizing problem.

4.1 Ali Baba and the Forty Thieves Algorithm

4.1.1 Basic AFT Algorithm

Ali Baba and the forty thieves (AFT) algorithm draws inspiration from the well-known narrative of AFT. Thanks to its straightforward and uncomplicated framework, this human-behavior-based algorithm can be conveniently implemented for different applications [1]. In this algorithm, the environment serves as the search space, the thieves represent search agents, and positions within the town stand in for potential solutions. Furthermore, Ali Baba's house is modeled as the objective function, with Ali Baba symbolizing the global solution.

The storyline narrates a pursuit-related scenario involving a forty-thief group who aim to capture Ali Baba to seek revenge and retrieve their stolen treasure [2]. Their search is conducted iteratively, in which solutions at a specific iteration are

Advanced Techniques for Optimal Sizing of Analog Integrated Circuits: Quantum Computing, Machine Learning, and Bio-inspired Optimization, First Edition.
Trang Hoang, Thinh Quang Do, Thang Quoc Nguyen, Hoang Trong Nguyen, Lihong Zhang, Octavia A. Dobre, and Trung Q. Duong.

generated based on previous iterations' results [3]. The AFT algorithm employs a population during its search process to represent the cooperative actions of the thieves. Meanwhile, within the storyline, the protagonist and Ali Baba's maid, Marjaneh, continuously takes countermeasures to hinder the gangs' efforts, increasing the complexity and difficulties of the search process. This might draw an analogy with the analog circuit sizing design, in which noise and higher complexity at advanced technology nodes can cause difficulties in the design procedure.

Since human social interaction is utilized to formulate the optimization model of the AFT, three main presumptions to be satisfied are as follows [4]:

- The forty thieves collaborate to obtain directions from a fellow thief or someone else to find Ali Baba's house. These directions might not always be accurate.
- The forty thieves travel a defined distance from their starting position to Ali Baba's house.
- By a certain proportion, Marjaneh can mislead the thieves and protect Ali Baba using her clever tricks multiple times.

Considering a population of n thieves, the search strategy of the thieves can be divided into three main cases:

Case 1: The position of the thieves when chasing Ali Baba using the direction they receive from a source is given by

$$x_i^{t+1} = gbest^t + [Td^t(best_i^t - y_i^t)r_1 + Td^t(y_i^t - m_{a(i)}^t)r_2]\ sgn(rand - 0.5); \quad p \geq 0.5, q > Pp^t, \tag{4.1}$$

where x_i^{t+1} is the current position of the ith thief, y_i^t stands for Ali Baba's position regarding the ith thief, $best_i^t$ denotes the best position obtained by the ith thief, $gbest^t$ identifies the global best position achieved by any thief, $m_{a(i)}^t$ stands for Marjaneh's intelligence level to conceal the ith thief, Td^t is the tracking distance of the thieves given by (4.2), Pp^t is the potential conceptual ability for Ali Baba given by (4.3), r_1, r_2, $rand$, p, and q are random values created within the range [0, 1], and $sgn(rand - 0.5)$ can assume values of -1 or 1 to control the quality of exploitation by changing the direction of the search.

The tracking distance parameter T_d, which characterizes the exploration process, gradually diminishes throughout the iterations. The decrease in T_d corresponds to local search in favorable regions of the search space around potentially best solutions, or in other words, the exploitation phase [3]. The T_d parameter is given by

$$T_d^t = \alpha_0 \times e^{-\alpha_1 (\frac{t}{T})^{\alpha_1}}, \tag{4.2}$$

where t and T represent the current and maximum iteration values, respectively, α_0 identifies an initial estimate of the tracking distance, and α_1 is a constant value to control exploitation and exploration.

On the other hand, the perception potential parameter P_p, employed to regulate the extent of exploitation, is continuously increased during the AFT's implementation. This helps the search process avoid being stuck in local optima, approach global optima, and quantify the quantity of exploitation through an in-depth search around the best solution [3, 5]. The formula for P_p is expressed as

$$P_p^t = \beta_0 \times \log\left[\beta_1\left(\frac{t}{T}\right)^{\beta_1}\right], \tag{4.3}$$

where β_0 is the final approximation of the probability that the thieves will achieve their goal at the end of the search process and β_1 is a constant number to control the capacity for exploration and exploitation.

Furthermore, the parameter a in $m_{a(i)}^t$ of (4.1) is defined as

$$a = [(n-1) \times rand(n, 1)], \tag{4.4}$$

where $rand(n, 1)$ is a vector of random numbers uniformly scattered within the interval [0,1] and n is the element number.

Additionally, based on the values of a predetermined objective function, Marjaneh modifies her intelligence level $m_{a(i)}^t$ if the quality of the thieves' introduced solution exceeds that of the previous ones. The formula for $m_{a(i)}^t$ is written as

$$m_{a(i)}^t = \begin{cases} x_i^t, & \text{if } f_{obj}(x_i^t) \text{ is more optimal than } f_{obj}(m_{a(i)}^t), \\ m_{a(i)}^t, & \text{otherwise,} \end{cases} \tag{4.5}$$

where $f_{obj}(.)$ represents the value of the objective function or fitness value.

Case 2: The thieves might find out that they have been deceived and begin to search for Ali Baba in a random manner, in which their positions can be obtained as

$$x_i^{t+1} = Td^t[(u_j - l_j) \times rand + l_j];\ p \geq 0.5, q \leq Pp^t, \tag{4.6}$$

where u_j and l_j indicate the respective lower and upper bound of the jth dimension of the search space.

Case 3: To further enhance the exploration as well as exploitation aspects of AFT, an alternative search strategy is developed to support the finding of positions beyond those of (4.1), which is given by

$$x_i^{t+1} = gbest^t - [Td^t(best_i^t - y_i^t)r_1 + Td^t(y_i^t - m_{a(i)}^t)r_2]sgn(rand - 0.5);\quad p < 0.5. \tag{4.7}$$

The summary and clarification of AFT's procedure are illustrated by the pseudocode of Algorithm 4.1.

Algorithm 4.1: Ali Baba and the Forty Thieves (AFT) Algorithm

Randomly initialize the position x of n thieves in the population;
Initialize $gbest$, $best_i$, $m^t_{a(i)}$;
Initialize $\alpha_0, \alpha_1, \beta_0, \beta_1$;
Evaluate the values of the objective function f_{obj} for each thief;
Set $t \leftarrow 1$;
while $t < T$ **do**
    Calculate Td^t, P^t_p;
    if $p \geq 0.5$ **then**
        if $q > P^t_p$ **then**
            Update the thieves' positions using (4.1) of **Case 1**;
        else
            Update the thieves' positions using (4.6) of **Case 2**;
    else
        Update the thieves' positions using (4.7) of **Case 3**;
    Update $gbest$, $best_i$;
    Update $m^t_{a(i)}$ according to (4.5);
    $t \leftarrow t + 1$
return $gbest$;

4.1.2 Self-Adaptive AFT (SaAFT)

According to Eq. (4.2), since α_0, α_1, and T are constants, at a specific iteration t, the value of the tracking distance T^t_d is the same for the whole population of thieves. This might degrade the efficiency of AFT since each search agent would demand different values of T^t_d. In fact, search agents that are trapped in local optima regions require a large tracking distance to explore the search space and escape from these undesirable areas, whereas other search agents utilize small tracking distances to exploit the favorably local areas [2]. As a result, it is necessary that T^t_d be adaptable and suitable for each search agent at each iteration [5], which motivated the enhanced version involving the adjustable tracking distance parameter

$T^t_{d_i}$ for each individual, termed as self-adaptive AFT (SaAFT). The parameter adjustment can also be advantageous for analog circuit sizing optimization regarding the prevention of premature convergence in large and complex design spaces.

SaAFT employs a fitness-based adjusting mechanism for the tracking distance T^t_d. When the fitness value of a thief does not change or worsen, indicating the situation of being stuck in local optima, a large tracking distance is reasonable to evade this region. Moreover, in the case of obtaining better values of the objective function, which means the possibility of approaching global optima increases, $T^t_{d_i}$ should be augmented to an appropriate extent to accelerate toward these optimal values and avoid excessive amounts that result in premature convergence.

In order to track the fitness results of the ith thief as an evaluation of AFT's individuals, cs^t_i and cf^t_i play the role as counters with the formula given by

$$cs^t_i = \begin{cases} cs^{t-1}_i + 1, & \text{if } f^t_{obj_i} \text{ is } \textbf{more} \text{ optimal than } f^{t-1}_{obj_i}, \\ 0, & \text{otherwise,} \end{cases} \tag{4.8}$$

$$cf^t_i = \begin{cases} cf^{t-1}_i + 1, & \text{if } f^t_{obj_i} \text{ is } \textbf{not} \text{ as optimal as } f^{t-1}_{obj_i}, \\ 0, & \text{otherwise,} \end{cases} \tag{4.9}$$

In terms of $T^t_{d_i}$, the two counters cs^t_i and cf^t_i together establish the condition for its adjusting mechanism as

$$T^t_{d_i} = \begin{cases} T^t_{d_i} \times r^t_i, & \text{if } (cs^t_i > \theta) \text{ or } (cf^t_i > \theta) \text{ or } (rand > \rho), \\ T^t_{d_i}, & \text{otherwise,} \end{cases} \tag{4.10}$$

where ρ is a constant to support the randomness of $T^t_{d_i}$ and θ is the threshold to control the tracking distance T_d via cs_i and cf_i.

The r^t_i parameter in (4.10), representing the adjusted value of $T^t_{d_i}$, is expressed as

$$r^t_i = \begin{cases} \frac{2}{c}, & \text{if } c < 1, \\ c, & \text{otherwise.} \end{cases} \tag{4.11}$$

In (4.11), c is the ratio between the tracking distance and perception potential of each thief at iteration t, which is given by

$$c = \left| \log\left(\frac{T^t_{d_i}}{P^t_{p_i}} \right) \right|. \tag{4.12}$$

The implementation of the SaAFT is further illustrated in Algorithm 4.2.

Algorithm 4.2: Self-Adaptive Ali Baba and the Forty Thieves (AFT) Algorithm

Randomly initialize the position x of n thieves in the population;
Initialize $gbest, best_i, m^t_{a(i)}$;
Initialize $\alpha_0, \alpha_1, \beta_0, \beta_1, cs^t_i, cf^t_i$;
Evaluate the values of the objective function f_{obj} of each thief;
Set $t \leftarrow 1$;
while $t < T$ **do**
    Calculate $T^t_{d_i}, P^t_{p_i}$;
    if $p \geq 0.5$ **then**
        if $q > P^t_p$ **then**
            Update the thieves' positions using (4.1) of **Case 1**;
        else
            Update the thieves' positions using (4.6) of **Case 2**;
    else
        Update the thieves' positions using (4.7) of **Case 3**;
    Calculate cs^t_i, cf^t_i;
    if $(cs^t_i > \theta)$ *or* $(cf^t_i > \theta)$ *or* $(rand > \rho)$ **then**
        Calculate r^t_i;
        Recalculate $T^t_{d_i} \leftarrow T^t_{d_i} \times r^t_i$;
    else
        Keep $T^t_{d_i}$;
    Update the thieves' positions based on **Case 1** or **Case 2** or **Case 3**;
    Update $gbest, best_i$;
    Update $m^t_{a(i)}$ according to (4.5);
    $t \leftarrow t + 1$
return $gbest$;

4.2 Drawer Algorithm

Drawer algorithm (DA), introduced by Trojovska and his colleagues in 2023, is primarily inspired by the process of choosing items from several drawers to make the best possible combination. In constructing DA, multiple drawers are assumed to be available, each holding a specific number of items. In fact, humans regularly attempt to grab their desired objects from the drawers of a closet and generate appropriate combinations to meet their needs. For instance, assume that a closet has various drawers, each of which contains only one type of item such as watches, jewelry, shoes, ties, hats, and so on. In this case, a person has to choose a suitable style involving all the mentioned things. This is one of the high-level behaviors of humans [6] that drives the motivation to formulate an optimization algorithm.

To mathematically model how people choose their favorable items out of the closet drawers, a DA's distinct features are presented as follows [6]:

- By utilizing random combinations of individuals of the population in the updating process, significant modifications are brought about for the individuals' positions. Consequently, stagnation, in which all members of the population are gathered at the same position, can be further averted, contributing to the algorithm's efficiency.
- The number of objects in the drawers reduces throughout the iterations of the algorithm. In the beginning, the number of objects is initialized at its maximum value, which creates more combinations of drawers with higher possibilities. This corresponds to the priority for exploration and global search in the solution space to determine major optimal regions. On the other hand, as the iterations of the optimization process increase, exploitation and local search are prioritized to further ensure the global optimum convergence. This is illustrated by the gradually decreasing number of objects inside the drawers and hence restricted movements in the individuals' positions. Thanks to DA's mentioned gradual alteration in the number of objects, an appropriate balance between exploration and exploitation can be maintained to possibly enhance the algorithm's performance.

Since DA is a meta-heuristic method that utilizes an iterative approach to optimization problems, at each iteration, its population searches the solution space in order to converge toward the best solution. Considering an n-individual or n-object population with D dimensions, at iteration t, the ith individual is represented by

$$x_i^t = \left\{x_{i,1}^t, x_{i,2}^t, \dots, x_{i,j}^t, \dots, x_{i,D}^t\right\},$$

where $x_{i,j}^t$ is the jth dimension of the ith individual of the population at iteration t.

In every iteration, based on a fitness function or objective function of the optimization problem, the individuals' fitness values are evaluated as

$$f_{obj}(x_i^t) = f_{obj}(x_{i,1}^t, x_{i,2}^t, \dots, x_{i,D}^t).$$

In DA, the basic idea for updating its population is to choose a carefully selected set of drawers containing variable information, which is demonstrated by the presence of a commode with the same number of drawers. The values for each drawer in this commode are determined by

$$d^t_{i,j} = x^t_{rand(n^t),j}, \tag{4.13}$$

where $d^t_{i,j}$ is the jth dimension of the ith individual of the selected drawer at iteration t, and $rand(n^t)$ denotes the random function to generate a uniformly random integer in the range $(0, n^t]$. In addition, n^t is the number of objects at iteration t and is expressed as

$$n^t = \left[\left(1 - \frac{t}{T}\right) \times n\right], \tag{4.14}$$

where T is the maximum iteration of the optimization process and $t \in [1, T]$. It is obvious from (4.14) that DA's number of objects is initialized at its maximum value and then gradually decreases as the iteration proceeds. This complies with the algorithm's mentioned unique characteristic, which is the initial priority for exploration and ultimate priority for exploitation, to balance between them for DA's best effectiveness. This method should be particularly suitable and reasonable for analog sizing optimization. Considering its large and complicated search space, exploration of the entire space needs to be prioritized to determine promising regions, which is followed by the exploitation of these areas for optimal dimensions of circuit components.

In terms of the update of individuals in DA's population, a random combination from the selected drawer $d^t_{i,j}$ is generated to orientate the individuals into the search space. This random combination can be formulated as

$$c^t_{i,j} = d^t_{j,rand(n^t)}, \tag{4.15}$$

where $c_{i,j}$ is the jth dimension to guide the ith individual of the population at iteration t.

Subsequent to calculating the random composition $c_{i,j}$, the algorithm's individuals are updated according to the equations given by

$$x_i^{t_{temp}} = \begin{cases} x_i^t + rand(0,1) \times (c_i^t - rand(2) \times x_i^t), & \text{if } f_{obj}(c_i^t) \text{ is more optimal than } f_{obj}(x_i^t), \\ x_i^t + rand(0,1) \times (x_i^t - x_{best}), & \text{otherwise,} \end{cases} \tag{4.16}$$

$$x_i^{t+1} = \begin{cases} x_i^{t_{temp}}, & \text{if } f_{obj}(x_i^{t_{temp}}) \text{ is more optimal than or as optimal as } f_{obj}(x_i^t), \\ x_i^t, & \text{otherwise,} \end{cases} \tag{4.17}$$

where $x_i^{t_{temp}}$ is the new status of the ith individual, x_{best} is the currently best individual of the population, $rand(0, 1)$ and $rand(2)$ are uniformly distributed random values created within the range $[0, 1]$ and $[0, 2]$, respectively.

To summarize, DA's procedure can be illustrated by the pseudocode of Algorithm 4.3.

Algorithm 4.3: Drawer Algorithm (DA)

Randomly initialize individuals x_i^t of the population;
Evaluate the values of the objective function f_{obj} for each individual;
Set $t \leftarrow 1$;
while $t < T$ **do**
  Calculate n^t and then $d_{i,j}^t$ according to (4.14), (4.13), respectively;
  Generate the random composition $c_{i,j}^t$ based on (4.15);
  Calculate the new status of the individuals $x_i^{t_{temp}}$ using (4.16);
  Update the individuals based on (4.17);
  Save the currently best individual x_{best};
  $t \leftarrow t + 1$
return x_{best};

4.3 Political Optimizer

Political optimizer (PO), a newly developed algorithm by Q. Askari et al., draws inspiration from different phases of the Western political process [7]. It is a human-behavior-based algorithm well-suited for real-world problems that challenge traditional gradient-based methods [8]. PO employs the analogy between the model of a political system and the behavior of politicians and the optimization process to implement the optimization algorithm [8].

In fact, politics involves the management and administration of a region, state, or country. Within this framework, party-based political systems, which mainly include one-party political systems, two-party political systems, dominant-party political systems, and multi-party political systems, play a vital role [9]. In addition, parliamentary systems or presidential systems also appear in the governance of other states or countries [10]. The generalization of PO can be established thanks to the integration of the common features of these systems, including [8]:

- Political parties and constituencies
- Collaboration among politicians within the same party
- Election within the party to select leader
- Competition through inter-party elections
- The analytics of candidates from the previous election
- Shifts in political affiliation

Table 4.1 Mapping between politics and optimization algorithm.

Political entity	Optimization algorithm mapping
Party member	Candidate solution in the population
Party member's goodwill	Position of the candidate solution in the search space
Political member's goodwill	Performance-related optimization parameters
Election process	Objective function
Number of votes obtained by a party member	Fitness of the candidate solution

- Pre-election campaign for votes during the election process
- Cooperation among elected officials in parliament
- Government formation and parliamentary affairs after winning elections.

Politics is itself an optimization process with regard to two aspects, namely (i) each individual desires to optimize their goodwill for electoral success and (ii) each party strives to secure the most parliament seats for government formation [7]. These two perspectives make politics inspirationally compatible with the formation of an optimization algorithm, based on the analogies illustrated in Table 4.1 [8].

The concept of utilizing politics as a model for an optimization algorithm is precedented. This is demonstrated by existing politics-inspired optimization algorithms such as greedy politics optimization (GPO) [11], motivated by political strategies used in state assembly elections; parliamentary optimization algorithm (POA) [12], driven by the competition among parliamentarians for leadership positions; and the election campaign optimization (ECO) algorithm [13], influenced by the motivation of candidates during election campaigns. However, the following characteristics set PO apart from those belonging to the politics-based group [8]:

- Politics-inspired algorithms such as GPO, POA, and ECO are related to solely one political phase. On the other hand, PO's mathematical model involves the multi-phased political process, including party formation and constituency allocation, election campaign, party switching, inter-party election, and parliamentary affairs.
- A unique updating mechanism for the candidate solutions' positions called recent past-based position updating strategy (RPPUS) is employed. Unlike other updating strategies with respect to the global best individual, RPPUS takes into account the position at the previous iteration of that candidate solution in the same way as the learning behavior of politicians from the previous election. Thanks to its consideration of the past solution and thereby determining the upcoming search areas, the decision whether to exploit the promising region

or eliminate the undesirable region to explore another can be implemented in a more feasible and logical manner. Therefore, the RPPUS update mechanism of PO should be beneficial to optimization problems with broad and complex search spaces, especially for determining optimal dimensions of circuit components in the analog sizing optimization process.

- Each candidate solution plays a dual role, specifically as a party member and an election candidate as well. As a result, each individual can update its position based on both the party leader and the constituency winner. Consequently, a parallel bidirectional search can be enabled with wider coverage across the solution space. This aspect should be helpful for the convergence speed and quality of optimal values for circuit parameters in analog sizing optimization.

PO is organized into five phases, namely party formation and constituency allocation, election campaign, party switching, inter-party elections, and parliamentary affairs. The first two phases occur only once during initialization, while the remaining four phases are executed in a continuous loop. The mathematical model that maps each phase of the multi-party political system onto PO's implementation is demonstrated in the pseudocode of Algorithm 4.4, which is further clarified in Sections 4.3.1–4.3.4. Furthermore, for the sake of clarity and comprehension, variables used in PO's formulation are illustrated in Table 4.2.

Table 4.2 Variables used in PO's formulation.

Variable	Description
n	Number of parties, constituencies, and members in each party
P	Whole population of political parties
P_i	ith political party
p_i^j	jth member of the ith party
$p_{i,k}^j$	$k|rmth$ dimension of the jth member of the ith party
d	Dimension of the jth member of the ith party
P^*	Set of party leaders
p_i^*	Leader of the ith political party
C	Set of all constituencies
C_j	jth constituency
C^*	Set of constituency winners
c_j^*	Winner of the jth constituency
$best$	Best member of the whole population
λ	Party switching rate
λ_{max}	Upper bound of the party switching rate λ
T	Maximum number of iterations

Algorithm 4.4: Political Optimizer (PO)

/***Initialization***/

Initialize n, λ_{max}, T;

Initialize P consisting of P_i and p_i^j;

Evaluate the values of the objective function f_{obj} of each member p_i^j;

Select P^*, C^* and *best* according to the calculated values of f_{obj};

$t \leftarrow 1, \lambda \leftarrow \lambda_{max}$;

$P(t-1) \leftarrow P(t)$;

$f_{obj}(P(t-1)) \leftarrow f_{obj}(P(t))$;

while $t < T$ **do**

  $P_{temp} \leftarrow P(t)$;

  $f_{obj}(P_{temp}) \leftarrow f_{obj}(P(t))$;

  for $i = 1 : n$ **do**

    for $j = 1 : n$ **do**

      $p_i^j(t) = \text{ElectionCampaign}(p_i^j(t), p_i^j(t-1), p_i^*, c_i^*)$;

  PartySwitching(P, λ);

  /***Election phase***/

  Evaluate the values of the objective function f_{obj} of each member p_i^j;

  Select P^*, C^* and *best* according to the calculated values of f_{obj};

  ParliamentaryAffairs(C^*, P);

  $P(t-1) \leftarrow P_{temp}$;

  $f_{obj}(P(t-1)) \leftarrow f_{obj}(P_{temp})$;

  $\lambda \leftarrow \lambda - \lambda_{max}/T$;

  $t \leftarrow t+1$

return *best*

4.3.1 Party Formation and Constituency Allocation

An n-party political population P can be demonstrated as

$$P = \{P_1, P_2, \dots, P_i, \dots, P_n\}.$$

Each party P_i is made up of n party members, which is illustrated by

$$P_i = \{p_i^1, p_i^2, \dots, p_i^j, \dots, p_i^n\}.$$

The d-dimensional party member p_i^j is represented as

$$p_i^j = [p_{i,1}^j, p_{i,2}^j, \dots, p_{i,k}^j, \dots, p_{i,d}^j].$$

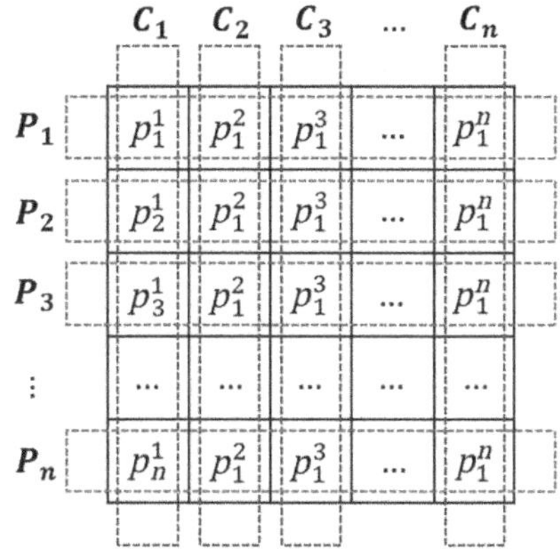

Figure 4.1 Logical division of the population P in terms of political parties and constituencies.

Along with the role of a party member, an individual can also act as an election candidate. It is supposed that there are n electoral districts, in this case, the set of all constituencies C can be represented as

$$C = \{C_1, C_2, \ldots, C_i, \ldots, C_n\}.$$

Assume that each constituency includes n members, then the ith constituency is given by

$$C_i = \left\{p_i^1, p_i^2, \ldots, p_i^j, \ldots, p_i^n\right\}.$$

Figure 4.1 demonstrates the systematic division of a political population into political parties and constituencies. In this scheme, the number of candidates in each party n equals the number of constituencies, which only requires the adjustment of a single parameter n.

The party leader is characterized as the fittest member or the member with the best objective value in a party. Similarly, a constituency winner, also called a member of parliament, is defined as the individual of the best fitness value in a specific constituency. The set of all party leaders P^* and constituency winners C^* are represented as

$$P^* = \{p_1^*, p_2^*, \ldots, p_i^*, \ldots, p_n^*\},$$
$$C^* = \{c_1^*, c_2^*, \ldots, c_i^*, \ldots, c_n^*\}.$$

4.3.2 Election Campaign

The purpose of the election campaign phase is to enhance the performance of election individuals, or in other words, to explore promising regions for solution candidates for better efficiency of PO. This phase consists of distinctive features that differentiate PO from other politics-based algorithms. These aspects include the RPPUS updating strategy as well as the dual updating role of political individuals with reference to both the party leader and the constituency winner. The election campaign phase can be mathematically mapped using (4.18) and (4.19) as

$$p_{i,k}^{j}(t+1) = \begin{cases} m^* + r(m^* - p_{i,k}^{j}(t)), & \text{if } p_{i,k}^{j}(t-1) \le p_{i,k}^{j}(t) \le m^* \text{ or } \\ & p_{i,k}^{j}(t-1) \ge p_{i,k}^{j}(t) \ge m^*, \\ m^* + (2r-1)|m^* - p_{i,k}^{j}(t)|, & \text{if } p_{i,k}^{j}(t-1) \le m^* \le p_{i,k}^{j}(t) \text{ or } \\ & p_{i,k}^{j}(t-1) \ge m^* \ge p_{i,k}^{j}(t), \\ m^* + (2r-1)|m^* - p_{i,k}^{j}(t-1)|, & \text{if } m^* \le p_{i,k}^{j}(t-1) \le p_{i,k}^{j}(t) \text{ or } \\ & m^* \ge p_{i,k}^{j}(t-1) \ge p_{i,k}^{j}(t), \end{cases} \tag{4.18}$$

$$p_{i,k}^{j}(t+1) = \begin{cases} m^* + (2r-1)|m^* - p_{i,k}^{j}(t)|, & \text{if } p_{i,k}^{j}(t-1) \le p_{i,k}^{j}(t) \le m^* \text{ or } \\ & p_{i,k}^{j}(t-1) \ge p_{i,k}^{j}(t) \ge m^*, \\ p_{i,k}^{j}(t-1) + r(p_{i,k}^{j}(t) - p_{i,k}^{j}(t-1)), & \text{if } p_{i,k}^{j}(t-1) \le m^* \le p_{i,k}^{j}(t) \text{ or } \\ & p_{i,k}^{j}(t-1) \ge m^* \ge p_{i,k}^{j}(t), \\ m^* + (2r-1)|m^* - p_{i,k}^{j}(t-1)|, & \text{if } m^* \le p_{i,k}^{j}(t-1) \le p_{i,k}^{j}(t) \text{ or } \\ & m^* \ge p_{i,k}^{j}(t-1) \ge p_{i,k}^{j}(t), \end{cases} \tag{4.19}$$

where r is a random number in the interval $[0, 1]$ and m^* first holds the value of kth dimension of the party leader $p_{i,k}^*$ then the constituency winner $c_{j,k}^*$.

The process of the election campaign can be further demonstrated by Algorithm 4.5. This pseudocode compares a candidate's previous fitness $f(p_i^j(t-1))$ with its current fitness $f(p_i^j(t))$ and determines whether to employ (4.18) or (4.19) accordingly. In both scenarios, the individual's position is initially updated based on the party leader p_i^*, followed by an update with respect to the constituency winner c_j^*.

4.3.3 Party Switching

Unlike real-life politics where the party-switching phase occurs simultaneously with its election campaign counterpart, the PO implements party switching after

Algorithm 4.5: ElectionCampaign $(p_i^j, p_i^j(t-1), p_i^*, c_j^*)$

if $f(p_i^j(t))$ *better than* $f(p_i^j(t-1))$ **then**
  for $k = 1 : d$ **do**
    $m^* \leftarrow p_{i,k}^*$;
    /*Position update with reference to the party leader*/
    Update $p_{i,k}^j(t)$ using (4.18);
    $m^* \leftarrow c_{j,k}^*$;
    /*Position update with reference to the constituency winner*/
    Update $p_{i,k}^j(t)$ using (4.18);
else
  for $k = 1 : d$ **do**
    /*Position update with reference to the party leader*/
    $m^* \leftarrow p_{i,k}^*$;
    Update $p_{i,k}^j(t)$ using (4.18);
    /*Position update with reference to the constituency winner*/
    $m^* \leftarrow c_{j,k}^*$;
    Update $p_{i,k}^j(t)$ using (4.18);
return $p_i^j(t+1)$

the election campaign. In this phase, every member p_i^j is chosen with a probability λ and changed to a random party P_r, in which p_i^j is swapped with the worst individual p_r^w of that P_r party. Algorithm 4.6 elaborates on the procedure for the party-switching phase.

Algorithm 4.6: PartySwitching (P, λ)

for $i = 1 : n$ **do**
  for $j = 1 : n$ **do**
    psr = random number in the interval[0; 1];
    if $psr < \lambda$ **then**
      r_i = random ***integer*** in the range[1; n];
      swap(p_r^w, p_i^j);

4.3.4 Parliamentary Affairs

Subsequent to the determination of all party leaders and constituency winners/parliamentarians from inter-party elections, the government is constructed. Each parliamentarian c_j^* adjusts its position according to a randomly selected parliamentarian c_r^*. In the case there is an improvement in the fitness of c_j^*, both the fitness and status of c_j^* are updated accordingly. This means that its corresponding p_i^j vector is also modified. The parliamentary affairs' operation can be further illustrated by Algorithm 4.7.

Algorithm 4.7: ParliamentaryAffairs (C^*, P)

for $i = 1 : n$ **do**
- r_i = random ***integer*** in the range$[1; n]$, $(r_i \neq j)$;
- a = random number in the range$[0; 1]$;
- $c_{new}^* = c_r^* + (2a - 1)|c_r^* - c_j^*|$;
- Evaluate the fitness value of c_{new}^*;
- **if** $f_{obj}(c_{new}^*)$ *is better than* $f_{obj}(c_j^*)$ **then**
  - $c_j^* \leftarrow c_{new}^*$;
  - $f_{obj}(c_j^*) \leftarrow f_{obj}(c_{new}^*)$;
  - $i \leftarrow$ party index of the winner of the jth constituency;
  - $p_i^j \leftarrow c_{new}^*$;
  - $f_{obj}(p_i^j) \leftarrow f_{obj}(c_{new}^*)$;

4.4 War Strategy Optimization

War strategy optimization (WSO), a recently developed metaheuristic optimization algorithm by Ayyarao et al. in 2022, draws inspiration from the three major phases of ancient war strategy, namely attack, defense, and reorganization. Besides the straightforward attack and defense strategy, reorganization involves the replacement or relocation of weak soldiers. In WSO, based on the determined strategy, the position of each soldier on the battlefield, or the position of each individual in the solution space, is dynamically updated toward that of the global optimum [14]. Furthermore, the so-called rank and weight of each soldier, which directly affects their movement, is adjusted according to the successful attack or the improvement in fitness values.

The inspiration for WSO can be demonstrated by ancient war strategy as follows. In ancient times, an army, normally consisting of numerous forces such as infantry, elephants, chariots, and so on, was essential to a kingdom's well-being and prosperity. It served as a means of prevention and protection against attacks by other dynasties. In a kingdom's army, war strategy, developed based on a mission's goals, potential threats, challenges, and opportunities, plays an indispensable part. It is regarded as a dynamic and adaptive process to deal with the progressively changing war conditions [15]. In the war strategy process, the king (or emperor) and the commanders of each unit direct the soldiers to organize their forces in a particular manner in order to reach the predetermined objectives and successfully accomplish the mission. This corresponds to the idea of an optimization problem and hence motivates the formulation of WSO, which is developed according to the following assumptions:

- The king is viewed as the army's supreme commander, while the commander is the soldier with the highest offensive capability. Therefore, each soldier has an equal chance of becoming either a commander or a king.
- The king's primary goal is to defeat the opposing leader, while the main objective for the soldiers is to attack the enemy and advance in rank.
- The position update of soldiers depends on those of both the king and commander. The flags on top of the king and commander indicate their positions, which are visible to all soldiers. Thanks to this update mechanism, candidate solutions can be adjusted according to both the best and second-to-best individual, enabling more options and directions to follow. Therefore, a higher chance of finding promising regions and high-quality optimal solutions, and hence increasing the convergence speed of the optimization process can be brought about. This might indicate the suitability of WSO for analog sizing optimization due to similar required outcomes for the optimization process.
- Soldiers are trained to follow strategic directions indicated by drumbeats or other musical signals under the king's instruction to adapt tactics based on the battlefield situation.
- If a military commander dies, the strategy is adjusted accordingly, requiring the remaining commanders to adapt and reorganize the war plan to maintain its progression.
- Soldiers with the weakest combat ability or those who are wounded are reorganized through replacement or relocation strategies.

The procedure for WSO in the pseudocode of Algorithm 4.8, consisting of four main phases namely attack strategy, rank and weight update, defense strategy, and replacement of weak soldier, is described in detail below.

Algorithm 4.8: War Strategy Optimization (WSO)

Initialize the size of the soldier population, the dimension of the war space (search space) and its lower bound and upper bound;
Initialize the positions of the king (K), commander (C) and the position of the soldiers;
Assign a value in the range [0,1] for ρ_r as the probability to determine the strategy to follow;
Evaluate the values of the objective function f_{obj} (fitness values, or attack strength) for each individual;
Arrange the attack strength of all soldiers in order and then determine the king and commander as the soldier with the best and second-to-best fitness, respectively;
Set $t \leftarrow 1$;
while $t < T$ **do**
    $\rho = rand$;
    if $\rho < \rho_r$ **then**
        Update the soldiers' positions based on (4.24) of the defense strategy to explore the war space;
    else
        Update the soldiers' positions based on (4.20) of the attack strategy to exploit the war space;
    Re-evaluate the attack strength of each individual and update their fitness according to (4.21);
    Update the rank and weight according to (4.22) and (4.23);
    Identify weak soldiers with the worst fitness and replace them according to (4.25);
    Update the positions of the king and commander;
    $t \leftarrow t + 1$
return K;

4.4.1 Attack Strategy

When the army follows the attack strategy, the king leads the army and exploits the opponent's weaknesses (promising areas of the search space) with continuous attacks. The king and commander move separately in two chariots marked with strategic flags on top to initiate a powerful offensive on the enemy forces. The soldiers' positions are updated according to those of their king and commander and the weight of the soldier himself, which is given by

$$x_i^{t+1} = x_i^t + 2 \times \rho \times (C - K) + rand \times (W_i \times K - x_i^t), \tag{4.20}$$

where x_i^{t+1} and x_i^t are the respective soldier's new position and previous position, ρ denotes the probability to choose between attack or defense strategy, W_i demonstrates the weight of the soldier, C and K represent the position of the king and commander, correspondingly.

According to (4.20), in the attacking phase at the start of the battle, troops spread out in multiple directions, making large moves to adjust their positions. Subsequently, the soldier's attacking force (or fitness value) is re-evaluated. When he successfully enhances his attack strength, his rank will improve, setting a positive example for others. On the other hand, when the new position is unsuitable for combat, the soldier returns to his previous position.

4.4.2 Rank and Weight Update

A soldier's rank is determined by his record of past success on the battlefield, which influences the weighing factor W_i. The rank indicates how close the soldier is to achieving the global optimum, indicating the progress of the optimization process toward the goal.

If the fitness value of the new position is not as good as that of the previous position, the soldier will return to its previous position, which can be expressed mathematically as

$$x_i^{t+1} = \begin{cases} x_i^{t+1}, & \text{if } f_{obj}(x_i^{t+1}) \text{ is more optimal than } f_{obj}(x_i^t), \\ x_i^t, & \text{otherwise,} \end{cases} \tag{4.21}$$

where $f_{obj}()$ is the value of the objective function or the fitness value of the individual to be considered.

In the case there is an improvement in the soldier's fitness value or a successful position update of the soldier, the rank R_i of the soldier will upgrade, which is written as

$$R_i = \begin{cases} R_i + 1, & \text{if } f_{obj}(x_i^{t+1}) \text{ is more optimal than } f_{obj}(x_i^t), \\ R_i, & \text{otherwise.} \end{cases} \tag{4.22}$$

Based on the R_i value obtained in (4.22), the weight W_i varies in an exponential manner relative to α, which is expressed as

$$W_i = W_i \times \left(1 - \frac{R_i}{T}\right)^{\alpha}, \tag{4.23}$$

where T is the maximum iteration of the optimization process and α is the index variation factor.

Equation (4.23) illustrates that the weights given to each soldier, which are adjustable by iteration, vary in large or small steps depending on the respective soldiers' steps. Moreover, a soldier's high fitness level corresponds to a lower weight, and vice versa.

4.4.3 Defense Strategy

In the defense strategy, the commander utilizes the army troops to arrange a chain formation around the king. This helps the whole army create better protection for the king to prevent losing the battle. In order to create a coordinated defense, each soldier adjusts their position based on the nearby soldiers' and the king's locations, which can be written as

$$x_i^{t+1} = x_i^t + 2 \times \rho \times (K - x_{rand}^t) + rand \times W_i \times (C - x_i^t), \tag{4.24}$$

where x_{rand}^t is the position of a random soldier at iteration t.

As shown in (4.24) and comparing with (4.20) of the attack strategy, its defense counterpart employs the position of a random soldier, which results in a higher level of exploration of the army troops (search agents) in the war field (search space) [14]. Furthermore, W_i directly affects the soldiers' position update, in which large W_i means a large step for position update and vice versa.

4.4.4 Replacement of Weak Soldiers

In each iteration, soldiers with the worst fitness are labeled as weak soldiers. Accordingly, the replacement approach involving substituting the weaker soldier with a randomly selected one is applied, as follows:

$$x_w^{t+1} = LB + rand \times (UB - LB), \tag{4.25}$$

where x_w^{t+1} is the new position of the weak soldiers, LB and UB are the lower bound and upper bound values of the soldiers' positions, respectively.

4.5 Case Study: Two-Stage Miller-Compensated Operational Amplifier with SaAFT and PO[1]

Operational amplifiers (op-amps), with the primary aim of amplifying voltage signals and providing a stable output, are considered one of the fundamental blocks in analog circuit design. They are integrated as sub-blocks in several different larger blocks with higher levels of complexity such as BGRs, low-dropout regulators, and analog-to-digital converters [16]. Depending on the applications and design specifications, the role of the op-amp differentiates; therefore, some performance metrics are prioritized at the expense of others [17]. With a view to designing an op-amp that can be virtually applied to different purposes, a balance among its universal design parameters is necessary while simultaneously keeping other parameters at acceptable levels. This corresponds to the purpose of the

1 This section has been published as part of [16].

multi-objective optimization, which should be the primary motivation to select the op-amp as a case study for the multi-objective optimization of analog circuit sizing.

The sections demonstrate the selection of objective function, accompanied by setup steps for the op-amp optimization problem as well as optimization results by SaAFT and PO. (Readers can examine further analysis for the two-stage Miller-compensated op-amp in [16].)

4.5.1 Choice of the Objective Function

As presented throughout Sections 2.1–2.3, an objective function (fitness function) is considered a standard to assess the individuals of the population and thereby exerts a direct influence on the evolutionary process. Therefore, the choice of a suitable objective function is crucial.

In the context of analog sizing optimization, the figure of merit (FoM), which typically refers to a quantitative measure to evaluate the performance of a system, can play the same role as the objective function. The FoM to compare the performance of the op-amp is given as [18]

$$\mathrm{FoM} = \frac{\mathrm{UGB} \times C_L}{I_{total}}, \tag{4.26}$$

where UGB is the unity gain-bandwidth product, C_L is the load capacitance at the output node, and I_{total} is the total current of the op-amp.

Although Eq. (4.26) evaluates the efficiency of the op-amp based on its unity gain-bandwidth product and the drivability of the load capacitance per unit current, the phase margin of the op-amp is overlooked. An op-amp with a larger value for the FoM is considered as being of better quality; however, an increase in C_L leads to a decrease in p_2, at the expense of the op-amp's phase margin, hence stability. In other words, considering (4.26), a large FoM value is no longer meaningful if the phase margin is too small [19], indicating the unsuitability of the above-mentioned objective function.

Taking into account the role of the op-amp's phase margin, the enhanced version of the FoM is expressed as [19]

$$\mathrm{FoM} = \frac{\mathrm{UGB} \times C_L}{I_{total}} \times \frac{tan(\mathrm{PM})}{tan(PM_{REF})}, \tag{4.27}$$

where PM_{REF} is the reference phase margin of the op-amp and chosen with the standard value of 60 deg [20].

4.5.2 Initialization Steps for Optimization

Regarding the two-stage Miller-compensated op-amp, its design is executed in the TSMCN65 process. The op-amp's setup condition includes $V_{DD} = 1.2$ V,

$I_{REF} = 20\ \mu A$, and $C_L = 1$ pF. The input common-mode voltage for both V_{inn} and V_{inp} is $V_{in_{CM}} = 650$ mV.

In total, there are ten variables in our optimization process, namely W_{01}, L_{01}, W_{23}, L_{23}, W_5, W_{47}, L_{457}, W_6, L_6, C_C, which are the values of MOSFETs' widths, lengths, and of capacitors within the circuitry. The specifications for the op-amp are as follows. First of all, every MOSFETs should operate in the saturation region with margins of 30 mV ensured for both the gate–source voltage V_{GS} and the overdrive voltage. For the sake of convenience, the mentioned condition is labeled $cond_1$. When $cond_1$ is satisfied, the standard for the op-amp's performance metrics ($cond_2$) is expressed as

$$A_V > 50\,\text{dB}, \tag{4.28a}$$

$$\text{UGB} > 50\,\text{MHz}, \tag{4.28b}$$

$$\text{PM} > 60^\circ, \tag{4.28c}$$

$$\text{CMRR} > 50\,\text{dB}, \tag{4.28d}$$

$$\text{Power} < 250\,\mu\text{W}, \tag{4.28e}$$

$$\text{SR} > 50\,\text{V}/\mu\text{s}. \tag{4.28f}$$

It should be noted that $cond_2$ is satisfied only when all design specifications of the op-amp above are met.

As a higher value of the FoM is preferable and $cond_1$ is prioritized over $cond_2$, the FoM can be rewritten with the addition of $cond_1$ and $cond_2$ as

$$\text{FoM} = \begin{cases} -1, & \text{if } cond_1 = 0, \\ 0, & \text{if } cond_1 = 1, cond_2 = 0, \\ \dfrac{\text{UGB} \times C_L}{I_{total}} \times \dfrac{tan(\text{PM})}{tan(60^\circ)}, & \text{otherwise,} \end{cases} \tag{4.29}$$

where the values of 1 and 0 are equivalent to each condition being satisfied and not satisfied, respectively.

When both $cond_1$ and $cond_2$ are not satisfied, the values of −1 and 0 are chosen for the FoM with the purpose of excluding the equivalent potential solutions. Moreover, the orders of −1 and 0 correspond to the priority of $cond_1$ and $cond_2$.

To ensure the saturation condition to the greatest possible extent, the bounds for optimization variables are determined as follows. Firstly, all MOSFETs' widths and lengths are initialized with the minimum values of the process node. Then, the widths and lengths of one MOSFET or one pair of symmetric MOSFETs are adjusted, while others might be kept constant or vary based on the basic knowledge of MOSFET's operation. This step is to find the temporary bounds for the optimization variables. Subsequently, this iterative process continues for the remaining MOSFET or pair of symmetric MOSFETs, in which the temporary bounds might be changed accordingly for better suitability. Finally, the bounds for optimization variables are obtained. We note that the purpose of this procedure is to ensure the saturation condition to limit the search space for better efficiency in finding the optimal FoM value.

As a result, the bounds for the optimization variables are obtained as

$$\begin{aligned}
&W_{01} \in [0.85\ \mu\text{m}, 4\ \mu\text{m}], L_{01} \in [0.23\ \mu\text{m}, 0.4\ \mu\text{m}],\\
&W_{23} \in [0.7\ \mu\text{m}, 1\ \mu\text{m}], L_{23} \in [0.06\ \mu\text{m}, 0.4\ \mu\text{m}],\\
&W_{47} \in [2\ \mu\text{m}, 2.8\ \mu\text{m}], W_5 \in [18\ \mu\text{m}, 23\ \mu\text{m}],\\
&L_{457} \in [0.1\ \mu\text{m}, 1\ \mu\text{m}], W_6 \in [16\ \mu\text{m}, 22\ \mu\text{m}],\\
&L_6 \in [0.25\ \mu\text{m}, 0.5\ \mu\text{m}], C_C \in [0.3\ \text{pF}, 1\ \text{pF}].
\end{aligned} \tag{4.30}$$

These bounds are obtained from the designer's experience and behave like rules of thumb to ensure control of the optimization process.

In essence, our optimization problem can be summarized as

$$\begin{aligned}
\mathit{maximize}\ \text{FoM}\ (&W_{01}, L_{01}, W_{23}, L_{23}, W_5, W_{47},\\
&L_{457}, W_6, L_6, C_C),
\end{aligned} \tag{4.31a}$$

$$s.t.\ V_{DD} = 1.2\ \text{V}, \tag{4.31b}$$

$$V_{in_{CM}} = 650\ \text{mV}, \tag{4.31c}$$

$$I_{REF} = 20\ \mu\text{A}, \tag{4.31d}$$

$$C_L = 1\ \text{pF}, \tag{4.31e}$$

$$(4.28), (4.30). \tag{4.31f}$$

4.5.3 Results and Discussion

This section presents the optimization results of the SaAFT and PO for the two-stage Miller-compensated op-amp circuit. For a fair comparison, both algorithms utilize a population of 16 individuals, in which each individual is a multi-dimensional variable containing the sub-variables in their decimal value range as in (4.30).

According to experimental results, regarding SaAFT, the optimal set of algorithm's parameters include $\alpha_0 = 1, \alpha_1 = 2, \beta_0 = 0.1, \beta_1 = 2, \theta = 10, p = 0.5$ while that of PO is given by $\lambda = \lambda_{max} = 0.8$. With all specifications in (4.28) satisfied, the FoM of 0.6034 and 0.6058 is obtained after 100 iterations for SaAFT and PO, respectively. The design parameters constituting the optimal FoM value are illustrated in Table 4.3.

Figure 4.2 illustrates the change in optimal FoM values over 100 optimization iterations for both SaAFT and PO. At the end of the optimization process, both algorithms achieve rather similar results for FoM of 0.6034 and 0.6057, respectively, which might indicate close-optimal solutions. Furthermore, it is worth taking into account the escape from the so-called "FoM-zero" region, where candidate solutions pass the saturation condition but do not meet specifications outlined in (4.28), as an indicator of the algorithm's capability in evading the local optima's areas. Indeed, SaAFT requires 8 iterations to break out of the undesirable "FoM-zero" region while PO only needs 1 iteration to complete the task. This efficiency might result from PO's unique RPPUS update strategy as well as the dual role of each solution candidate representing both a party member and an election candidate, as presented in Section 4.3.

A closer look at Figure 4.2 reveals that as the FoM value becomes positive, SaAFT speeds up considerably in its FoM upward trend, possibly thanks to the utilization of dynamic tracking distance T_d and perception potential parameter P_p. This is denoted by SaAFT's surpassing the FoM milestone of 0.55 at around the 30th iterations, which outperform that of PO by approximately 25 iterations. However, while the optimization process progresses toward completion, PO exhibits better convergence speed by achieving the FoM value of 0.6 at the 75th iteration, which means 20 iterations from the 0.55 checkpoint compared to the two-fold number 40 iterations by SaAFT. Subsequently, the FoM graph from SaAFT remains constant while that of PO experiences a slight increase, reaching their peak at 0.6034 and 0.6058, respectively.

Table 4.3 Optimum design parameters achieved for the two-stage Miller-compensated op-amp with their corresponding specifications.

Design parameters	Specification	SaAFT's optimum value	PO's optimum value
A_V (dB)	>50	51.05	50.02
UGB (MHz)	>50	68.77	66.81
PM (°)	>60	68.83	68.66
CMRR (dB)	>50	53.51	53.22
Power (μW)	<250	205.09	195.6
SR (V/μs)	>50	61.35	62.1

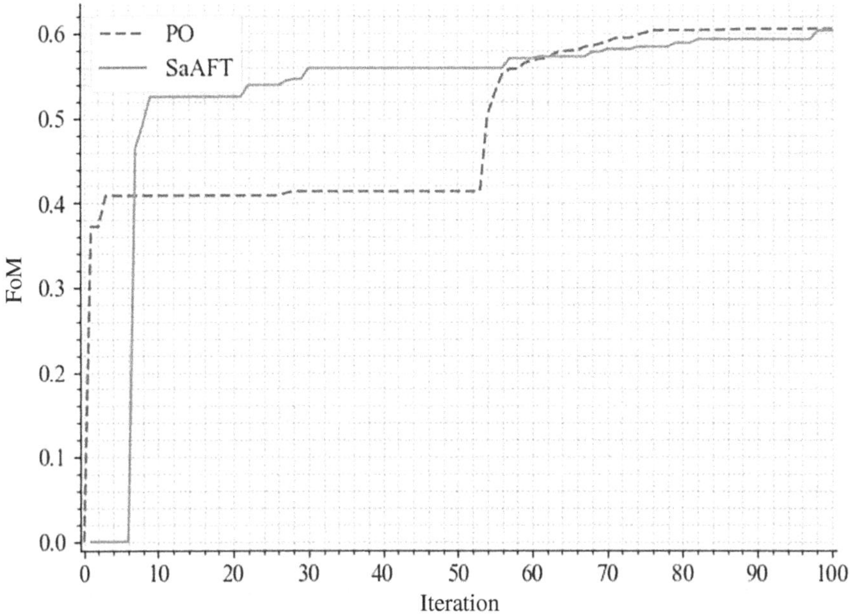

Figure 4.2 FoM convergence behavior comparison between SaAFT and PO over 100 iterations.

In addition, it is essential that the 5 nm process grid of the TSMCN65 process be considered when finalizing the numerical results for optimization variables of SaAFT and PO, which means that the widths and lengths of the MOSFETs after optimization need to be adjusted to compatible values with the process grid. Post-optimization dimensions of all MOSFETs in the circuit are provided in Table 4.4, while Table 4.5 demonstrates a comparison in performance between the two-stage Miller-compensated op-amp proposed in this section and other studies.

Based on Table 4.5, compared to [21], it is obvious that our work dominates in terms of FoM, A_V, PM, and $CMRR$. The higher power consumption versus that of [21] should be reasonable because of a greater amount of current for a two-fold A_V value. Regarding [22], its output capacitor C_L is ten times as large as that used in our work, which greatly influences its better FoM. Nevertheless, a larger current is required to drive the large C_L, resulting in a power consumption of almost one and a half times as much as that of our work. The mentioned scenario should be considered similar to that of [22] in terms of the connection between output capacitor C_L and power consumption.

Table 4.4 Post-optimization sizing results of optimized two-stage Miller-compensated op-amps.

Device	SaAFT's optimized size (W/L)	PO's optimized size (W/L)
M_0, M_1	2.21 μm/0.33 μm	2.29 μm/0.39 μm
M_2, M_3	1 μm/0.11 μm	0.99 μm/0.1 μm
M_4	2.73 μm/0.5 μm	2.52 μm/0.81 μm
M_5	20.51 μm/0.5 μm	18 μm/0.81 μm
M_6	21.13 μm/0.25 μm	22 μm/0.25 μm
M_7	2.73 μm/0.5 μm	2.52 μm/0.81 μm
C_C	0.31 pF	0.3 pF

Table 4.5 Performance summary of the proposed two-stage op-amp versus those of other studies.

Performance metrics	[21]	[22]	[23]	[24]	This work's SaAFT	This work's PO
CMOS process (nm)	65	65	180	180	65	65
Supply voltage (V)	1.1	1.1	1.8	1.8	1.2	1.2
C_L (pF)	0.2	10	10	1	1	1
A_V (dB)	21.6	43	75.62	75.35	51.05	50.02
UGB (MHz)	169.7	28	16.78	45.11	68.77	66.81
PM (°)	62.4	60	62.48	61.24	68.83	68.66
CMRR (dB)	35.5	N.A	N.A	N.A	53.51	53.22
Power (μW)	89	321	220	123.156	205.09	195.6
SR (V/μs)	288	N.A	15.88	N.A	61.35	62.1
FoM (HzF/A)	0.1432	0.9595	1.5214	0.6936	0.6034	0.6058

Compared with [23] and [24], owing to the larger range for the transistors' lengths as well as the higher process node's supply voltage, our work's A_V is approximately 25 dB lower. However, *UGB*, as well as *PM* performance, exceeds those of these two studies. Considering [24], despite our work's better *UGB* and *PM*, lower FoM and higher power dissipation are observed, mostly due to the difference in technology node (65 versus 180 nm) and lower supply voltage (1.2 versus 1.8 V). To summarize, although there might be competitiveness instead of pure dominance between FoM and design parameters between other research of different setup conditions and technology nodes, post-optimization results from our research still demonstrate its effectiveness as well as feasibility throughout the mentioned insights and comparisons.

References

1 B. Ozak, "Fast Convergent Ali Baba and Forty Thieves Algorithm for Inverse Kinematic Solution of 7-DOF Robotic Manipulator," *Int. J. Pioneer. Technol. Eng.*, vol. 2, no. 2, pp. 80536–80560, 2023.

2 M. Braik, "Enhanced Ali Baba and the Forty Thieves Algorithm for Feature Selection," *Neural Comput. Appl.*, vol. 35, pp. 1–32, 2022.

3 K. U. Rehman, I. Sajid, S.-D. Lu, S. Ahmad, H.-D. Liu, F. I. Bakhsh, M. Tariq, A. Sarwar, and C.-H. Lin, "Fast Tracking of Maximum Power in a Shaded Photovoltaic System Using Ali Baba and the Forty Thieves (AFT) Algorithm," *Processes*, vol. 11, no. 10, p. 2946, 2023.

4 P. Sharma, S. Thangavel, S. Raju, and B. R. Prusty, "Parameter Estimation of Solar PV Using Ali Baba and Forty Thieves Optimization Technique," *Math. Probl. Eng.*, vol. 2022, no. 1, p. 5013146, 2022.

5 M. Braik, M. Ryalat, and H. Alzoubi, "A Novel Meta-Heuristic Algorithm for Solving Numerical Optimization Problems: Ali Baba and the Forty Thieves," *Neural Comput. Appl.*, vol. 34, no. 1, pp. 409–455, 2022.

6 E. Trojovská, M. Dehghani, and V. Leiva, "Drawer Algorithm: A New Meta-heuristic Approach for Solving Optimization Problems in Engineering," *Biomimetics*, vol. 8, no. 2, p. 239, 2023.

7 A. Zhu, Z. Gu, C. Hu, J. Niu, C. Xu, and Z. Li, "Political Optimizer with Interpolation Strategy for Global Optimization," *PLoS One*, vol. 16, no. 5, p. e0251204, 2021.

8 Q. Askari, I. Younas, and M. Saeed, "Political Optimizer: A Novel Socio-Inspired Meta-Heuristic for Global Optimization," *Knowledge-Based Syst.*, vol. 195, p. 105709, 2020.

9 G. Sartori and P. Mair, *Parties and Party Systems: A Framework for Analysis, ser*, ECPR Press Classics. ECPR Press, 2005.

10 C. Menard and M. M. Shirley, *Handbook of New Institutional Economics.* Springer, 2008.

11 J. L. Melvix, "Greedy politics optimization: Metaheuristic inspired by political strategies adopted during state assembly elections," in *2014 IEEE International Advance Computing Conference (IACC)*, Gurgaon, India, 2014, pp. 1157–1162.

12 A. Borji, "A new global optimization algorithm inspired by parliamentary political competitions," in *MICAI 2007: Advances in Artificial Intelligence*, vol. 4827, Nov. 2007, pp. 61–71.

13 W. Lv, C. He, D. Li, S. Cheng, S. Luo, and X. Zhang, "Election Campaign Optimization Algorithm," *Procedia Comput. Sci.*, vol. 1, no. 1, pp. 1377–1386, 2010.

14 T. S. L. V. Ayyarao, N. S. S. Ramakrishna, R. M. Elavarasan, N. Polumahanthi, M. Rambabu, G. Saini, B. Khan, and B. Alatas, "War Strategy Optimization Algorithm: A New Effective Metaheuristic Algorithm for Global Optimization," *IEEE Access*, vol. 10, pp. 25073–25105, 2022.

15 V. R. Gatla, S. K. Injeti, S. Kotte, P. Kumar Polamarasetty, R. S. S. Nuvvula, A. S. Singh Vardhan, M. Singh, and B. Khan, "An Effective Approach for Extracting the Parameters of Solar PV Models Using the Chaotic War Strategy Optimization Algorithm with Modified Newton Rapson Method," *IEEE J. Electron Devices Soc.*, vol. 12, pp. 849–858, 2024.

16 T. Q. Do, H. T. Nguyen, B. D. E. McNiven, T. Hoang, L. Zhang, D. A. Octavia, and T. Q. Duong, "An Auto-Adjusting Hybrid Quantum Genetic Algorithm-Spectre Platform for the Multi-Objective Optimization of Analog Circuit Sizing," *Alexandria Eng. J.*, vol. 116, pp. 574–586, Mar. 2025.

17 M.V. Harsha and B. P. Harish, "An integrated MaxFit genetic algorithm-SPICE framework for 2-stage op-amp design automation," in *Proc. IEEE Comput. Soc. Ann. Symp. VLSI (ISVLSI)*, Hong Kong, China, Jul. 2018, pp. 170–174.

18 W. Sansen, *Analog Design Essentials*. Berlin, Germany: Springer, 2006.

19 H. Yoshizawa, "An Improved Figure-of-Merit Equation for Op-Amp Evaluation," *IEICE Electron. Exp.*, vol. 12, p. 20150533, Aug. 2015.

20 B. Razavi, *Design of Analog CMOS Integrated Circuits*, 2nd ed. New York, United States: McGraw-Hill, 2017.

21 R. Rashid and N. Nambath, "Area Optimisation of Two Stage Miller Compensated Op-Amp in 65 nm Using Hybrid PSO," *IEEE Trans. Circuits Syst. II Exp. Briefs*, vol. 69, no. 1, pp. 199–203, Jan. 2022.

22 M. El-Sisi and M. Dessouky, "Procedural analog design automation using building block optimization," in *Proc. IEEE International Conference on Electronics, Circuits, and Systems (ICECS)*, Cairo, Egypt, Dec. 2015, pp. 474–477.

23 P. Das and B. Jajodia, "Design automation of two-stage operational amplifier using multi-objective genetic algorithm and SPICE framework," in *Proc. Int. Conf. Invent. Comput. Technol. (ICICT)*, Nepal, Jul. 2022, pp. 166–170.

24 S. A. Abdelaal, A. Hussein, and H. Mostafa, "A Bayesian optimization framework for analog circuits optimization," in *Proc. 15th Int. Conf. Comput. Eng. Syst. (ICCES)*, Cairo, Egypt, Dec. 2020, pp. 1–4.

5

Machine Learning in Analog Circuit Sizing Optimization

This chapter provides a systematic perspective on using machine learning (ML) in circuit sizing optimization. ML is a heuristic method compared to deterministic methods described in Chapters 2–4. Among various ML models, it is challenging to select the optimal solution to your design problem. Even with the appropriate model, it often relies on the designers' experience to perform efficient training due to complex design spaces and large amounts of hyperparameters. Moreover, circuit datasets for ML are not always available and redundant enough, and in some cases are heavily dependent on certain applications. Therefore, this chapter will point out the suitable ML models for circuit sizing specifically with benefits and drawbacks, while always observing the current situations of the designers.

5.1 Machine Learning Overview

With the advancement of IC technologies, the need for ML has become increasingly evident. Data are generated from numerous sources at an unprecedented rate. This explosion of data, referred to as big data, presents both opportunities and challenges. The sheer volume and complexity of data make it nearly impossible for humans to process and analyze it manually. This is where ML steps in, offering automated methods to efficiently handle, analyze, and extract valuable insights from big data.

5.1.1 Machine Learning: A Heuristic Sandbox

One of the critical reasons for the growing need for ML is the necessity to process large amounts of data rapidly. Humans, despite their cognitive abilities, quickly become overwhelmed and exhausted when faced with massive datasets. This leads to errors, delays, and inconsistencies in data analysis. On the other hand,

Advanced Techniques for Optimal Sizing of Analog Integrated Circuits: Quantum Computing, Machine Learning, and Bio-inspired Optimization, First Edition.
Trang Hoang, Thinh Quang Do, Thang Quoc Nguyen, Hoang Trong Nguyen, Lihong Zhang, Octavia A. Dobre, and Trung Q. Duong.

machines can process data at incredible speeds, performing complex calculations and analyses in a fraction of the time. Its ability to consistently and objectively extract features from data can both remove the biased behavior of humans and uncover hidden patterns and insights that might be overlooked by human analysts. Despite initially lacking intelligence, ML can eventually handle tasks that would be impractical or impossible for humans to perform on their own due to massive amounts of data.

Time reduction is another significant factor driving the adoption of ML. In today's fast-paced world, timely decision-making is crucial across various domains. Machines, equipped with ML algorithms, can analyze data in real time and provide actionable insights almost instantaneously. This rapid processing capability not only enhances efficiency but also enables organizations to respond quickly to changing conditions, make informed decisions without unnecessary delays, and achieve reasonable time-to-market.

However, merely having the ability to compute fast is not enough. Machines need to approximate how humans make decisions through self-conscious evaluation. This involves learning from data to make precise predictions and decisions. ML models achieve this by identifying patterns and relationships within the data. This process of approximation is akin to how humans make decisions based on their experiences and knowledge. By mimicking this process, machines can provide valuable insights and recommendations that aid in decision-making.

To achieve a good approximation, ML is not a single-step process but involves a continuous learning or training period, during which algorithms are fed large datasets and iteratively adjust their parameters (weights) to improve accuracy and performance. This learning process is crucial for developing models that can generalize well to new, unseen data and provide reliable predictions. Just as humans refine their skills through practice, machines enhance their capabilities through continuous learning and adaptation, which is considered a heuristic technique.

ML models can be generalized as a sandbox with the following components:

- **Input dataset**: Most ML models require input data for training. Features or patterns from these data are extracted by ML algorithms to provide solutions for various problems such as classification and prediction. These input data are often collected from casual environments and may contain noise or other environmental impacts. Additionally, the data can be heterogeneous in terms of shape and dimensions, potentially causing incompatibility with ML algorithms. Therefore, pre-processing methods for input data are crucial, as they can significantly affect the overall performance of ML models.
- **Algorithms and parameters**: ML algorithms are employed to extract valuable information from input data. These algorithms utilize numerous internal parameters that are gradually tuned to optimize data processing. Furthermore,

other parameters not directly related to the data but crucial for training, such as learning rates and the number of layers, are configurable variables known as hyperparameters. The entire set of parameters is critical for the success of the training process and often requires substantial designers' expertise.

- **Output evaluation and hyperparameter tuning**: After partial training, ML models are prepared to process more new data. However, it is important to evaluate the current performance of these models to ensure their accuracy and ability to generalize to unseen data. Additionally, hyperparameter tuning is equally important in the training process. Although hyperparameters do not store the pattern memory of the data, they significantly impact the results of ML models in terms of convergence speed and loss reduction. This is because hyper-parameters represent various configurations or scenarios within the system.

Within the scope of this book, we are pursuing methods and techniques to aid the electronic automation design (EDA) process, and ML stands out as one of the most effective ones. While IC development paved the way to enable the ML revolution, now it seeks a fair return. Datasets for circuits are not widely available though, and the process of applying ML for EDA can be even more challenging and complicated. For the rest of this chapter, we are looking for suitable ML models and how to use them wisely for the problem of analog circuit sizing. Please notice that methods and algorithms in Chapters 2–4 are deterministic, while ML is known to be heuristic, and proper navigation for ML training can be confusing and incomprehensible for even the trainers.

5.1.2 Procedure of Applying Machine Learning

ML involves several critical steps, each essential for developing models that can effectively learn from data and make accurate predictions (Figure 5.1). The process begins with data collection, where relevant data are gathered from various sources. These data are then cleaned and pre-processed to ensure it is suitable for analysis. Feature selection and extraction may follow, where the most relevant attributes of the data are identified. The dataset is then divided into training and testing sets, and an appropriate ML algorithm is chosen for feature extraction. The ML model is trained using the training data, and its performance is verified using the testing data. Finally, the model is fine-tuned and optimized before being deployed for real-world applications.

Training and testing are fundamental components of the ML process. During the training phase, the model learns from the training data by adjusting its parameters to minimize error. This involves iteratively processing the data and updating the model based on the feedback it receives. Once the model is trained, it is evaluated

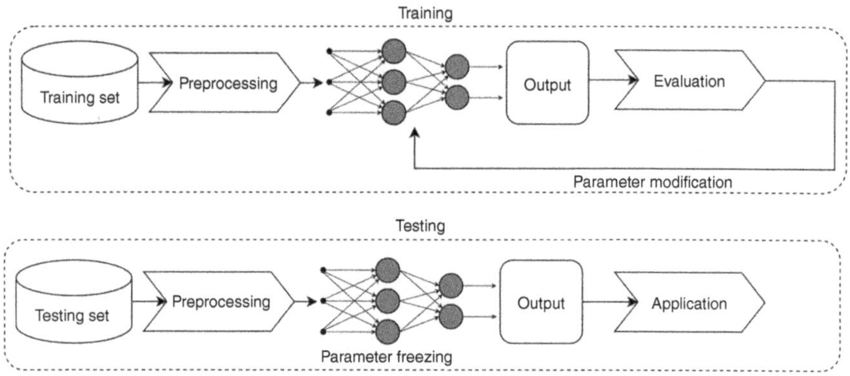

Figure 5.1 Procedure of training and testing for a general ML model.

using the testing data, which it has not seen before. This evaluation helps assess the model's capability to generalize to new, unseen data. The goal is to ensure that the model performs well on not only the training data but also the testing data, indicating its robustness and reliability.

ML problems such as overfitting, convergence issues, and gradient explosion can hinder the development of effective models. Overfitting occurs when a model learns the training data too well, including its noise and outliers, resulting in poor generalization of new data. Convergence issues arise when the model fails to find the optimal solution during training, often due to improper parameter settings or algorithmic limitations. Gradient explosion, particularly in deep learning, refers to the exponential growth of gradients during backpropagation, leading to instability and divergence. These problems highlight the need for appropriate algorithms and techniques to ensure the model trains effectively and generalizes well.

ML instability underscores the importance of a careful setup and human expertise. Instability can result from various factors, including data quality issues, improper algorithm selection, and inadequate parameter tuning. To mitigate these risks, practitioners must meticulously set up the ML pipeline, ensuring that each step is carefully executed and validated. Human experience and domain knowledge play a crucial role in this process, as they guide the selection of appropriate algorithms, parameter settings, and validation techniques. By combining rigorous technical practices with expert insights, the stability and reliability of ML models can be significantly enhanced.

In EDA, datasets of circuits are generally not available, due to differences in formats, applications, and confidentiality. As a result, organizations are required to perform numerous simulations to gather enough data to form a dataset. This poses a threat to the aim of objective learning of ML models, and designers may choose slower methods that are less reliant on the dataset. Let us discuss some of the popular ML models in Sections 5.1.3–5.1.5, where we describe in detail what and how to apply ML to specific problems and tasks and indicate how suitable they can be used for analog circuit sizing automation.

5.1.3 Supervised Learning

One of the three fundamental ML models is supervised learning, which makes use of labeled datasets to classify or predict outputs based on the corresponding inputs. This type of learning involves a supervisor who provides the correct output for each training example. The primary objective is for the model to learn the mapping from inputs to outputs so that it can accurately predict the output for new, unseen data. Supervised learning encompasses a variety of tasks, including classification and regression, which are applied across numerous domains such as finance, healthcare, and marketing.

In supervised learning, the training dataset consists of input–output pairs, where the input is a data vector and the output is the target variable. For classification tasks, the target variable represents distinct categories or classes. For example, in a fraud detection system, actions are classified as either "fraudulent" or "normal." The model learns to distinguish between these categories based on the features extracted from the actions, such as executing time or fraudster information. In regression tasks, the target variable is continuous, such as predicting house prices based on features such as square footage, number of bedrooms, and location. During training, supervised algorithms iteratively adjust their parameters to minimize the error between the predicted and actual outputs. This process is guided by a loss function, which quantifies the difference between the predicted and true values.

One of the main challenges in supervised learning is overfitting. To deal with this, techniques such as cross-validation, regularization, and pruning are employed. Cross-validation involves partitioning the training data into subsets and training the model on each subset while validating it on the remaining data. Regularization adds a penalty for complexity to the loss function, discouraging the model from becoming too intricate. Pruning simplifies decision trees by removing branches that contribute little to the model's performance.

Within supervised learning, two primary tasks, as mentioned above, are classification and regression. Both tasks involve learning a mapping from input features to output targets, but they differ in the nature of these targets and the specific objectives of the models.

Classification is the task of predicting discrete class labels for given input data. The goal is to categorize data points into one of several predefined classes. Classification models are evaluated based on their ability to assign these labels correctly, and common algorithms include logistic regression, decision trees, random forests, support vector machines, and neural networks. The performance of classification models is typically measured using metrics such as accuracy, precision, recall, and the F1 score.

Regression, on the other hand, deals with predicting continuous numerical values. The objective is to model the relationship between input features and a continuous output variable. Regression models aim to minimize the difference between the predicted and actual values, often using metrics such as mean squared error, mean absolute error, and R-squared to evaluate their performance. Common regression algorithms include linear regression, polynomial regression, decision trees, random forests, and neural networks. While both classification and regression are supervised learning tasks, they have distinct differences in their approach and application. Classification focuses on categorizing inputs into discrete classes, making it suitable for problems where the output is a label or category. Regression, conversely, is used for problems where the output is a continuous value, making it ideal for forecasting and predicting numerical outcomes.

For the tasks within analog circuit sizing procedure, supervised learning may have limited usage due to the lack of data amount and proper applications. These ML models, however, can be oriented to be a scoreboard for faster performance evaluation in replacement of time-consuming simulation tools. To achieve that, the supervised models are required to be pre-trained with the help of these simulation tools, to learn how to judge generated circuits or propose the nominal values of sizing parameters. As ML is heuristic, designers may face the challenge of uncertain, time-consuming effort, while being limited by the specific circuitry.

5.1.4 Unsupervised Learning

Unsupervised learning, in comparison to supervised learning, attempts to analyze unlabeled datasets and cluster data into subsets (or clusters). In other words, unsupervised learning provides a means for uncovering hidden patterns, structures, and relationships within the data, facilitating a deeper understanding of the dataset and enabling data-driven insights.

Clustering is one of the most prominent tasks in unsupervised learning, aiming at grouping a set of objects in such a way that objects in the same group (or cluster) are more similar to each other than to those in other groups. This similarity is typically measured using distance metrics such as Euclidean distance, Manhattan distance, or cosine similarity. Clustering algorithms are used extensively in various applications:

- **k-means clustering** [1]: This algorithm partitions the dataset into k distinct clusters by minimizing the variance within each cluster. It is widely used thanks to its simplicity and efficiency, although it requires the number of clusters (k) to be specified in advance.

- **Hierarchical clustering**: This method builds a tree of clusters, which can be either agglomerative (bottom-up) or divisive (top-down). It provides a visual representation called a dendrogram and does not require the number of clusters to be predefined.
- **Density-based spatial clustering of applications with noise** [2]: This algorithm identifies clusters based on the density of data points, making it effective for datasets with noise and varying cluster sizes. It is particularly useful for spatial data and anomaly detection.

Dimensionality reduction is another critical task in unsupervised learning, focusing on reducing the number of random variables under consideration by obtaining a set of principal variables. This is essential for managing high-dimensional datasets, where too many features can lead to the "curse of dimensionality" and make the analysis computationally intensive and less effective. The key techniques include:

- **Principal component analysis** [3]: This technique transforms the original features into a smaller set of uncorrelated variables called principal components, which capture the most variance in the data. It is widely used for feature reduction, noise reduction, and data visualization.
- **t-distributed stochastic neighbor embedding** [4]: This is a non-linear dimensionality reduction technique particularly well-suited for visualizing high-dimensional data in two or three dimensions. It captures local relationships between data points and is often used in exploratory data analysis.

By reducing the dimensionality, these techniques help in simplifying models, reducing overfitting, and improving computational efficiency. They also make it easier to visualize and interpret complex datasets, facilitating better decision-making.

Association rule learning is a method for discovering interesting relations between variables in large datasets. It is particularly useful in transactional data analysis, such as market basket analysis, where the goal is to identify sets of items that frequently co-occur in transactions. The key algorithms include:

- **Apriori algorithm**: This algorithm identifies frequent item sets and generates association rules by exploring the dataset iteratively, increasing the size of item sets by one at each step.
- **Equivalence class clustering and bottom-up lattice traversal**: This algorithm uses a depth-first search strategy to find frequent item sets, making it faster and more efficient for large datasets than Apriori.

Similar to supervised learning, the use of unsupervised learning in analog circuit sizing can be vague and problematic. The most useful application can be clustering generated circuits into groups of different qualities and providing meaningful feedback for the generative models to focus on interesting groups. Insufficient data for training can also lead to inconsistent performance of training and testing, so designers can set unsupervised learning aside as being a risky heuristic method.

5.1.5 Reinforcement Learning

Reinforcement learning (RL) refers to the third set of ML methods, which learns without sample data but using trial-and-error processes. Based on experience from previous successive learning steps, RL models can improve the correctness of their policies (how they determine the action) and achieve the solution with the highest reward for the given problems, which explains "reinforcement" in its name.

In other words, RL improves its *reward* by selecting a proper sequence of *actions*, which was controlled by a *policy*. For the use in analog circuit sizing, actions can mean modification to the sizing parameters within the circuit, and the environment for these actions to be executed is the circuit itself. The reward is the performance of the circuit, which is returned to the agents (cores of RL) to produce better actions for the following steps. In the RL scenario, a single action makes no major affection on the environment, which means these models aim for better long-term rewards. The intermediate reward in the middle of an episode may be low, such as taking a run-up for a momentum gain before a race, to achieve a higher reward in some points in the future. As a result, RL often takes intensive iterations to complete, since it needs to learn the best route to the final goal. To accomplish this, exploitation and exploration are usually considered. Exploration refers to the behavior of trying to execute new actions regardless of the current state, while exploitation means the action is selected based on the highest value of reward obtained in previous iterations after selecting that action.

The primary components of RL include the agent, environment, states, actions, and reward (Figure 5.2). The agent is the learner or decision-maker who seeks to

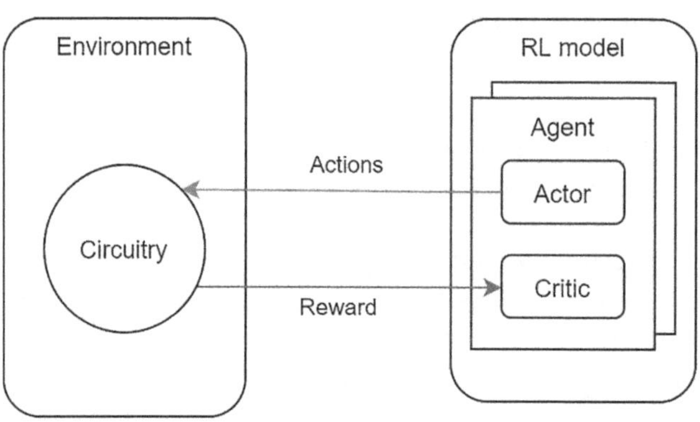

Figure 5.2 Illustration for operation of an RL model.

achieve a specific goal by taking action. The environment represents the external system with which the agent interacts. It provides feedback to the agent based on the actions taken. States represent the current situation or configuration of the environment. Each state is a snapshot of the environment at a particular time, containing all the relevant information needed for decision-making. Actions are the set of all possible moves the agent can make in a given state. The choice of action influences the state of the environment, potentially leading to a new state. Rewards are the feedback signals received by the agent from the environment after taking an action. They serve as a measure of the success or failure of the agent's actions. The primary objective of the agent is to maximize the cumulative reward over time. This involves learning an optimal policy, which is a strategy or mapping from states to actions that maximize the expected reward.

RL algorithms can be broadly categorized into three types: value-based, policy-based, and model-based methods. Value-based algorithms focus on estimating the value of states or state-action pairs. The value represents the expected cumulative reward starting from a given state or state-action pair. Examples of value-based algorithms include Q-learning and deep Q-networks. These algorithms aim to learn the optimal value function, which can then be used to derive the optimal policy. Policy-based algorithms, on the other hand, directly learn the policy without explicitly estimating value functions. These algorithms optimize the policy to maximize the expected reward. Examples include REINFORCE and proximal policy optimization (PPO). Policy-based methods are particularly useful in environments with large or continuous action spaces, where value-based methods might struggle. Model-based algorithms involve learning a model of the environment's dynamics. This model predicts the next state and reward given the current state and action. The agent uses this model to simulate future states and rewards, enabling planning and decision-making. Examples of model-based algorithms include model-based RL and Dyna-Q. These algorithms can be more sample-efficient than value-based or policy-based methods, as they leverage the learned model to explore and optimize the policy.

Despite its promising potential, RL faces several challenges. One significant issue is the difficulty in acknowledging and understanding every piece of information from the environment. In many real-world scenarios, which include analog circuit sizing, the environment can be highly complex and dynamic, with numerous states and variables that are difficult to capture accurately. This complexity makes it challenging for a single agent to learn an optimal policy effectively. To address this issue, multi-agent systems are often employed. In multi-agent RL, multiple agents interact with the environment and with each other. Each agent may focus on different aspects of the environment or collaborate to achieve a common goal. By sharing information and strategies, these agents can better explore and exploit the environment, which leads to improved performance and robustness. For example, in robotic swarm systems, multiple robots can work together to complete tasks more efficiently than a single robot could on its own. Multi-agent systems, however, can also introduce new challenges, such as coordination and communication among agents. Ensuring that agents work together effectively and do not interfere with each other's learning processes requires careful design and tuning of the algorithms. However, the potential benefits of multi-agent systems in dealing with complex environments make them a valuable approach to RL.

Another significant challenge in RL is handling multi-objective reward functions. The circuit sizing problem is one of those with many objectives, and RL agents need to consider multiple objectives or criteria simultaneously. For instance, an agent must balance power, speed, linearity, and other figures of merit. Achieving a single-objective reward is relatively straightforward, but optimizing for multiple conflicting objectives is much more complex. Multi-objective RL aims to address this challenge by learning policies that can balance multiple objectives. One approach is to combine the different objectives into a single scalar reward using a weighted sum or other aggregation methods. However, this approach requires carefully tuning the weights, which can be difficult and may not capture the true trade-offs between objectives. Another approach is to use Pareto optimization, where the goal is to find a set of Pareto-optimal solutions. These solutions represent different trade-offs between the objectives, allowing decision-makers to choose the most appropriate one based on their preferences. However, finding and maintaining a diverse set of Pareto-optimal solutions can be computationally expensive and challenging.

RL has proven to be more useful for analog circuit sizing since it requires no dataset and gradually tackles the complexity of the circuits. Its stability, however, can be extremely difficult to achieve if either the circuits are too complex, or the agents cannot find an optimal route to reach the required performance.

5.2 Neural Networks

Neural network is a branch of ML, while deep learning is a branch of neural networks. Going down the ladder of exploring the human brain, researchers have been looking for conventional methods to imitate this main functional organ, which is essential for us to think, to learn, and to be. By this fundamental idea, neural network creates a set of processing nodes with numerous links among them, just the same as brain neurons are connected to each other through the axons. This explains the name of "neural networks," which some other documents may use the name "perceptrons."

Neural networks are used in almost any kind of ML model, no matter if it belongs to the category of supervised, unsupervised, or RL. Various neural network architectures can be designed to tackle different problems. For example, convolutional neural networks are widely used for image processing tasks thanks to their ability to automatically and hierarchically learn spatial hierarchies of features from raw image data. Convolutional layers of these networks can apply filters to the input data, detecting features such as edges, textures, and patterns. These features are then pooled and passed through fully connected layers to make predictions. Convolutional neural networks have been highly successful in tasks such as image classification, object detection, and image generation.

Recurrent neural networks, on the other hand, are designed to handle sequential data, making them suitable for tasks such as natural language processing (NLP), time series forecasting, and speech recognition. Recurrent neural networks have connections that loop back on themselves, allowing information to persist and be used in subsequent time steps. This feedback mechanism enables recurrent neural networks to maintain a memory of previous inputs, which is crucial for understanding context and dependencies in sequences. Variants of recurrent neural networks, such as long short-term memory networks and gated recurrent units, have been developed to address the problem of vanishing gradients and improve the ability to capture long-term dependencies.

Another significant advancement in neural network architectures is the development of transformers, which have revolutionized NLP tasks. Transformers use self-attention mechanisms to weigh the importance of different words in a sentence, capturing contextual relationships more effectively than traditional recurrent neural networks. This architecture has led to the creation of powerful models such as bidirectional encoder representations from transformers (BERT) and generative pre-trained transformers, which have set new benchmarks in various NLP tasks, including text generation, translation, and comprehension. These transformers are usually referred to as large language models. A pre-trained large language model for IC design can be extremely useful, though it requires large datasets of circuits which have not been available yet. Nevertheless, it can assist designers in error detection, decision making, and other qualitative tasks. This trend is sure to accelerate in the near future, as large language model models become more and more intelligent.

Neural networks have also been extended to unsupervised learning and generative tasks. Generative adversarial networks, for instance, consist of two networks — a generator and a discriminator — that compete against each other. The generator creates fake data samples, while the discriminator tries to distinguish between real and fake samples. Through this adversarial process, generative adversarial networks can generate realistic data, such as images, music, and even human-like text. Variational autoencoders are another type of generative model that learns to encode data into a latent space and then decode it back to reconstruct the original data, allowing for efficient data generation and manipulation. As mentioned in Sections 5.1.3 and 5.1.4 of supervised and unsupervised learning, generative models can produce circuits in batches to be evaluated later by another ML model. Being trained in the right tracks, generative models can hence overcome expensive and tedious simulation tools.

Despite these capabilities, neural networks also come with challenges. They require large amounts of labeled data for training, which can be expensive and time-consuming to obtain. Additionally, neural networks are computationally intensive, often requiring specialized hardware such as graphic processing units for efficient training and inference. The interpretability of neural networks is another concern, as their complex and layered structure can make it difficult to understand how they make decisions, leading to the term "sandbox" models. Researchers are actively working on developing techniques to make neural networks more interpretable and trustworthy, such as attention mechanisms and feature attribution methods.

5.3 Hyperparameter Search

Hyperparameters refer to the parameters that must be determined before the start of the training process. These parameters do not directly affect the learning ability of ML models, but determine how other parameters are tuned during training, or in other words, determine the convergence and efficiency of the ML algorithms. Examples of hyperparameters include the learning rate and the number of layers in a neural network, the regularization parameter in a regression model, or the discount rate in RL models.

The process of hyperparameter tuning is essential because the choice of hyperparameters can significantly impact the model's ability to learn from the data and generalize to unseen data. Proper tuning can lead to substantial improvements in model accuracy, efficiency, and robustness. On the other hand, poor hyperparameter choices can result in underfitting, where the model is too simple to capture the underlying patterns in the data, or overfitting, where the model is too complex and captures noise in the data, leading to poor generalization. Therefore, finding the optimal set of hyperparameters is crucial for building effective ML models.

Several techniques can be employed for hyperparameter tuning. One of the simplest and most commonly used methods is grid search. Grid search involves defining a set of possible values for each hyperparameter and exhaustively evaluating the model performance for every possible combination of these values. Although this method is straightforward to implement, it can be computationally expensive and time-consuming, especially when dealing with a large number of hyperparameters or a large search space. Moreover, grid search treats each hyperparameter independently, which may not be efficient if there are interactions between hyperparameters.

Another popular method for hyperparameter tuning is random search. Unlike grid search, which evaluates all possible combinations, random search samples a fixed number of hyperparameter combinations randomly from the predefined search space. This method is more efficient than grid search because it does not require evaluating all combinations, making it feasible to explore larger search spaces. Studies have shown that random search can often find better hyperparameter settings in less time compared to grid search, especially when only a few hyperparameters significantly impact model performance.

Advanced techniques for hyperparameter tuning include Bayesian optimization, which uses a probabilistic model to predict the performance of different hyperparameter settings and guides the search process toward promising regions of the search space. Bayesian optimization balances exploration and exploitation by focusing on areas that are likely to yield high performance while still exploring new regions. This method is more sophisticated and can be more efficient than grid or random search, particularly for complex models with many hyperparameters.

The choice of the evaluation metric is another important aspect of hyperparameter tuning. The performance of different hyperparameter settings is typically assessed using cross-validation, which involves partitioning the data into training and validation sets multiple times and averaging the results. This helps to ensure that the evaluation is robust and not overly dependent on a particular data split. Common evaluation metrics include accuracy, precision, recall, F1-score for classification tasks, and mean squared error or mean absolute error for regression tasks. The choice of metric should align with the specific goals and requirements of the ML application.

Circuit sizing is a complex problem to solve, with many objectives to achieve. Finding a Pareto front for these objectives with a vast volume of design space can seriously hinder the progress of efficient ML training. Hyperparameter tuning for these cases can be vital for the whole training process. Designers can implement some of the methods described above, or rely on previous experience to execute the tuning. Nevertheless, the process can be rigorously challenging, similar to other heuristic methods.

5.4 Deep Reinforcement Learning

Deep reinforcement learning (DRL) is a subfield of ML that combines the decision-making capabilities of RL with the powerful representation learning of deep neural networks. In traditional RL, an agent learns to take actions in an environment to maximize cumulative rewards through trial and error, often relying on tabular methods or simpler function approximations. However, these approaches struggle to handle high-dimensional state or action spaces, such as raw image inputs or complex environments. DRL addresses this limitation by using deep neural networks to approximate the value functions, policies, or models of the environment, enabling the agent to process and act on large, unstructured data effectively.

One of the most well-known examples of DRL is the deep Q-network algorithm, which combines Q-learning with convolutional neural networks to play video games directly from pixel data. DRL has since been extended to a wide range of algorithms, such as policy-based methods like PPO and actor-critic frameworks like deep deterministic policy gradient. These methods allow DRL agents to tackle complex tasks in robotics, autonomous driving, finance, and beyond. However, DRL also introduces challenges, such as increased computational requirements, sensitivity to hyperparameters, and stability issues during training. Despite these challenges, DRL continues to be a powerful and growing field, advancing the boundaries of what machines can learn and achieve autonomously.

DRL has shown promise in tackling the sizing problem in analog circuit design, where the goal is to determine the optimal values for circuit parameters such as transistor widths, lengths, and bias currents to meet performance specifications. Sizing is a high-dimensional, non-linear optimization task with complex trade-offs between objectives such as power consumption, speed, gain, and noise. DRL addresses this challenge by treating it as a sequential decision-making problem, where the agent incrementally adjusts parameter values and receives a reward based on the circuit's performance metrics obtained through simulation. By learning a policy that maps circuit configurations to parameter adjustments, DRL can efficiently explore the parameter space and identify optimal or near-optimal solutions.

One key advantage of using DRL for the sizing problem is its ability to handle conflicting objectives and constraints. Traditional optimization methods often require predefined weights for each objective in a cost function, which may not accurately reflect the designer's priorities or trade-offs. DRL, however, can learn these trade-offs dynamically through multi-objective reward functions, allowing for a more flexible and adaptive approach to sizing. Furthermore, DRL agents can generalize their learned policies across similar circuits or design scenarios, enabling faster convergence in subsequent tasks. This reusability makes DRL particularly appealing for iterative design workflows, where circuit parameters need frequent refinement as specifications evolve. Despite challenges such as computational demands and the need for robust reward engineering, DRL offers a powerful tool to enhance the efficiency and innovation of the analog circuit sizing process.

5.5 Case Study: High-Level Design of Delta–Sigma Analog-to-Digital Converter[1]

5.5.1 Delta–Sigma ADC: Topology and Sizing

Analog-to-digital converters (ADCs) serve as bridges between natural analog signals and the digital realm, where most modern electronic devices function. Certain applications demand highly accurate digital representations of analog signals, necessitating high-resolution ADCs. Among them, Delta–Sigma ADCs, or Δ–Σ ADCs stand out for their ability to provide high-precision outputs through oversampling and noise shaping, though this comes at the expense of increased latency compared to other ADC types. However, as designers push for greater accuracy and complexity, these ADCs become prone to instability, despite their suitability for applications like audio processing and medical treatments. The modulator, a crucial component of any Δ–Σ ADC, largely determines the device's performance and architecture, as it defines both the signal transfer function (STF) and noise transfer function (NTF). Consequently, extensive research has focused on optimizing Δ–Σ modulator topologies to enhance stability and improve signal-to-noise ratio (SNR).

Designing a Δ–Σ modulator typically involves three key steps, akin to other analog circuit designs: topology synthesis, circuit sizing, and layout generation. The main focus is usually placed on topology and parameter sizing, as these factors primarily determine the final product's quality, while layout implementation mainly aims to minimize undesirable physical effects, such as parasitic capacitances and resistances. Δ–Σ ADCs often encounter stability issues when designers push for higher accuracy and performance, which also increases complexity [6], even when the design is intended for applications like audio processing or medical treatment. Several standard topology structures, such as cascaded integrators with feedback (CIFB), cascaded resonators with feedback (CRFB), cascaded integrators with feed-forward (CIFF), and cascaded integrators with distributed input and summed feed-forward (CIDIFF), have been proposed to achieve good performance in Δ–Σ modulators [7, 8].

Circuit sizing for this type of ADC requires careful consideration of topology synthesis, as this step directly impacts the performance of the final circuit. In this section, we present an example of complex circuits within a high-level context where both sizing and topology must be optimized simultaneously. As a result, the calculation space becomes vast, posing significant challenges for conventional optimization algorithms. RL, as mentioned earlier, appears to be particularly well-suited for such designs. First, RL optimizes the circuit

1 This section has been published as part of [5].

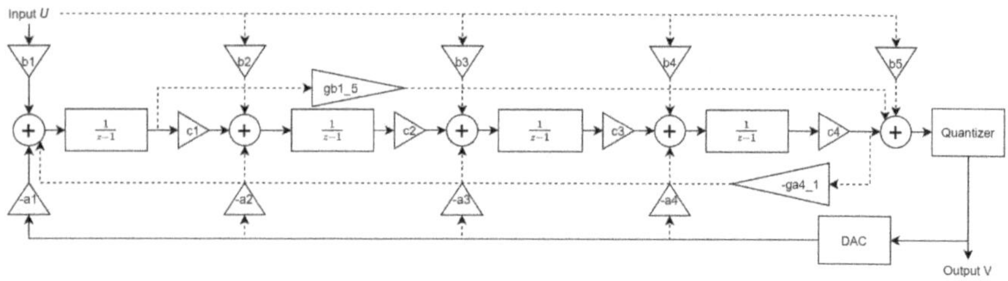

Figure 5.3 General topology of a fourth-order Δ−Σ modulator.

topology incrementally, providing stable updates, which makes it less sensitive to hyperparameter misconfigurations-an issue that often arises in circuit design due to epistemic uncertainty. Second, high-dimensional data, such as in this case, can be efficiently represented through the appropriate encoding of the broad action spaces, which can be further refined by using stochastic or deterministic policies.

A Δ−Σ modulator consists of a series of integrators, with the number of integrators corresponding to the modulator's order, along with a quantizer that converts the analog output signals into digital levels. The relationship between the input signal, noise, and transfer functions is expressed as

$$V(z) = STF(z)U(z) + NTF(z)E(z), \tag{5.1}$$

where $V(z)$ is the output after being processed by the quantizer, and $U(z)$ and $E(z)$ represent the input signal and the error introduced by quantization, respectively. All terms are expressed in the z-domain, with an integrator transfer function given by $\frac{1}{z-1}$, which introduces a delay. The general topology of a fourth-order Δ−Σ modulator is shown in Figure 5.3, where the output from the quantizer is processed through a digital-to-analog converter (DAC) to represent the corresponding voltage level in the analog domain, providing the feedback. In addition to the core components (integrators, DAC, and quantizer), several connections are formed between these elements to create distinct topologies with unique transfer functions. Forward signals from the input to the integrators are labeled as b_i, while signals from the feedback DAC are denoted by a_i. Internal connections between integrators are represented by their starting points, endpoints, and directions, using notation such as ga_{ij} or gb_{ij}.

5.5.2 Multi-agent Proximal Policy Optimization

PPO is one of the most widely used RL algorithms. As an on-policy algorithm, PPO relies on stable updates of agents through policy gradient optimization. To maintain stability and prevent erratic updates, the clipping mechanism was introduced, which sets boundaries on how much the gradient can be propagated back to adjust the weights. Additionally, PPO utilizes two neural networks: an actor to generate actions based on agent policies and a critic to evaluate and update those actions. While PPO is known for its stability, it typically takes longer to converge. In circuit design, however, time is not as critical as the efficiency of the final product. Furthermore, PPO agents may struggle to learn effectively in complex action spaces, as stochastic policies are less efficient at exploring large spaces, which could also delay convergence.

Although PPO is often considered less sample-efficient than other off-policy algorithms, making it unsuitable for multi-agent scenarios, this belief has been challenged. In [9], it was demonstrated that multi-agent PPO models can still yield competitive results, even when considering sample efficiency. Homogeneous agents, or identical agents, generally perform better than heterogeneous agents, which involve different types of actions corresponding to each agent. In the latter case, the hyperparameter settings become crucial in determining both the convergence rate and success.

Built on this, our proposed model incorporates two agents: one for topology construction and the other for sizing optimization. The topology construction agent has a set of three possible actions: disconnection (D), connection (C), and no change (R). The sizing optimization agent also has three actions: no change (R), increment by step size (I), and decrement by step size (D). After both agents select their respective actions, the multi-agent PPO model combines them into a joint action to be executed in the environment. The pseudo-code for our algorithm is provided in Algorithm 5.1. Training is carried out in batches, with the trajectory τ being stored in a buffer to provide data to both the agents and the environment. The shared observation space, denoted by s_t, is used to calculate the value function, which estimates the corresponding joint action a_t^j. Actions from the actors are generated based on separate observation spaces. The normalized advantage function $\hat{A}$ and the reward-to-go $\hat{R}_n$ are used in the loss functions $L(\theta)$ for the actors and $L(\phi)$ for the critics.

The observation space provided to the critic is the ABCD matrix, which represents the current topology. This space is shared, allowing both agents to access it and understand the impact of each other's actions. However, rewards are assigned different values based on the α coefficients, where each agent determines the relative importance of various objectives from its own perspective.

5.5.3 Results and Discussion

Figure 5.4 illustrates the convergence process of the SNR, showing a significant increase during the initial stages of the simulation. After 20 million steps, the value stabilizes with little further increase, ultimately reaching 102 dB, which represents a solid 10 dB margin.

Algorithm 5.1: Multi-agent PPO Algorithm for *Delta–Sigma* Modulator Topology Synthesis and Sizing

Initialize all hyperparameter settings;
while *finish* $\neq$ *True* **do**
  Initialize data buffer D and $batch\_id = 0$;
  while $batch\_\ id < batch\_size$ **do**
    $\tau \leftarrow \{\}$;
    Initialize states for two actors $h^{(1)}_{0,\pi}, h^{(2)}_{0,\pi}$;
    Initialize states for two critics $h^{(1)}_{0,\mathbf{V}}, h^{(2)}_{0,\mathbf{V}}$;
    for $n = 1$ **to** T **do**
      foreach *agent* $m \in \{1, 2\}$ **do**
        $a^{(m)}_t, h^{(m)}_{t,\pi} \leftarrow \pi(o^{(m)}_t, h^{(m)}_{t-1,\pi}; \theta^{(m)})$;
        $a^j_t \leftarrow \text{joint}(a^{(1)}_t, a^{(2)}_t)$;
        $v^{(m)}_t, h^{(m)}_{t,\mathbf{V}} \leftarrow \mathbf{V}(s_t, h^{(m)}_{t-1,\mathbf{V}}; \phi^{(m)})$;
      $r_t, s_{t+1}, o^{(1)}_{t+1}, o^{(2)}_{t+1} \leftarrow \text{step}(a^j_t; \mathbf{s})$;
      $\tau \mathrel{+}= [s_t, o^{(1,2)}_t, h^{(1,2)}_{t,\pi}, h^{(1,2)}_{t,\mathbf{V}}, a^j_t, r_t, s_{t+1}, o^{(1,2)}_{t+1}]$;
    $\hat{A} \leftarrow \text{GAE}(\tau)$;
    $\hat{R_n} \leftarrow \sum r_t$;
    Push data into D as data chunks;
    $batch\_id \mathrel{+}= 1$;
  Gather data from D;
  $\theta^{(1)} \leftarrow \text{Adam}(L(\theta^{(1)})), \theta^{(2)} \leftarrow \text{Adam}(L(\theta^{(2)}))$;
  $\phi^{(1)} \leftarrow \text{Adam}(L(\phi^{(1)})), \phi^{(2)} \leftarrow \text{Adam}(L(\phi^{(2)}))$;

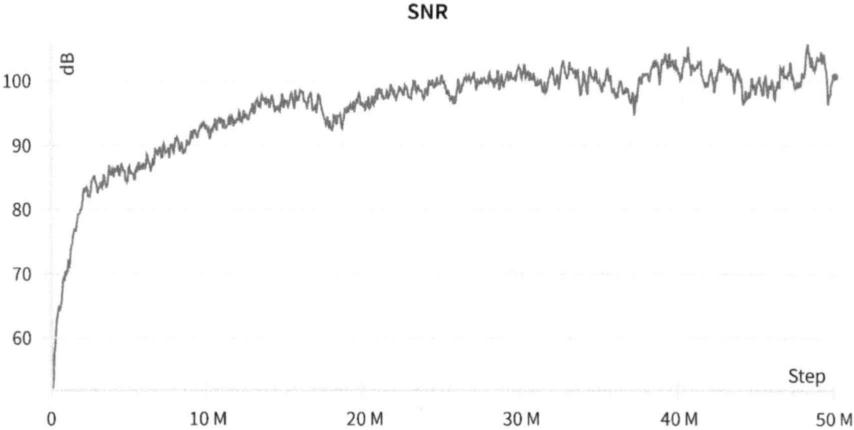

Figure 5.4 Convergence behavior of SNR.

Instead of relying solely on the actor's decision, we introduced 50 roll-out threads to perform parallel runs and select the most appropriate action based on the reward. While this approach could theoretically reduce effectiveness for long-term goals, the actor might end up in poor topology structures, causing fluctuations in the SNR trends as the simulation progressed. As shown in Figure 5.5, with only one roll-out thread, the model struggled to converge, fluctuating between 60 and 80 dB. However, by using a higher number of roll-out threads, our model achieved more stable convergence steps for the SNR.

Additionally, we examined the impact of step size and the initial starting point on model convergence. As shown in Figure 5.6, a large step size (higher step

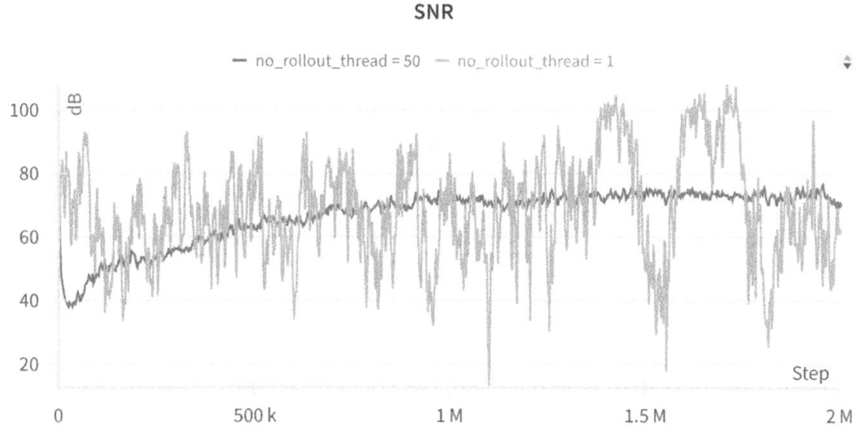

Figure 5.5 Impact of rollout thread on convergence of the proposed model.

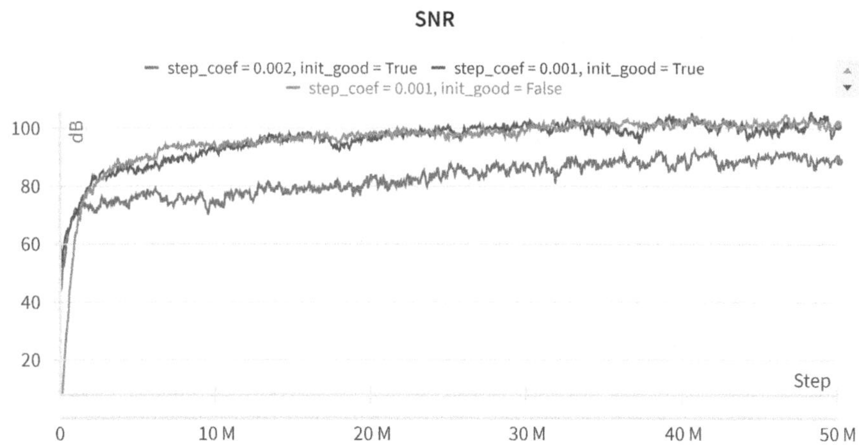

Figure 5.6 Impact of step size and initial starting topology on the convergence of proposed model based on SNR.

coefficient) can negatively affect the overall performance of the training. This is because it leads to poor actions that generate large gradients, resulting in significantly lower SNR compared to models with smaller step sizes. Regarding the initial starting points, we found that they had a minimal effect on the final convergence. While good initial structures may reach 80 dB earlier than poor ones, the convergence slope barely increases after this milestone.

References

1 K. P. Sinaga and M.-S. Yang, "Unsupervised K-Means Clustering Algorithm," *IEEE Access*, vol. 8, pp. 80716–80727, Apr. 2020.

2 E. Schubert, J. Sander, M. Ester, H. P. Kriegel, and X. Xu, "DBSCAN Revisited, Revisited: Why and How you Should (Still) Use DBSCAN," *ACM Trans. Database Syst.*, vol. 42, no. 3, pp. 1–21, Jul. 2017.

3 I. Jolliffe, *Principal Component Analysis*. Berlin, Heidelberg: Springer, 2011, pp. 1094–1096.

4 V. Marx, "Seeing Data as T-SNE and UMAP Do," *Nat. Methods*, vol. 21, pp. 930–933, May 2024.

5 T. Q. Do, L. Zhang, O. A. Dobre, T. Hoang, and T. Q. Duong, "Topology and parameter optimization of high-order Δ-Σ modulators towards superior efficiency and stability with multi-agent reinforcement learning," in *Proc. IEEE Intl. Symp. Circuits Syst. (ISCAS)*, London, UK, Jan. 2025.

6 R. Baird and T. Fiez, "Stability Analysis of High-Order Delta-Sigma Modulation for ADC's," *IEEE Trans. Circuits Syst. II*, vol. 41, no. 1, pp. 59–62, Jan. 1994.

7 R. Schreier and G. C. Temes, *Understanding Delta-Sigma Data Converters*, 1st ed. Wiley-IEEE Press, 2004.

8 J. M. De la Rosa and R. Del Rio, *CMOS Sigma-Delta Converters: Practical Design Guide*. John Wiley & Sons, 2013.

9 C. Yu, A. Velu, E. Vinitsky, J. Gao, Y. Wang, A. Bayen, and Y. Wu, "The surprising effectiveness of PPO in cooperative multi-agent games," in *Proc. Advances in Neural Inf. Proc. Systs.*, New Orleans, LA, Nov. 2022, pp. 24611–24624.

6

Quantum Mechanics and Analog Circuit Sizing Optimization

Quantum physics refers to the subatomic level behavior of physical entities. Endowed with two important characteristics of superposition and entanglement, quantum computing is promising to solve numerous complex problems, including circuit sizing design. At the moment, the use of quantum computing within the IC design field is minimal, on account of limited real quantum hardware, and the lack of necessary knowledge to operate quantum computing. This chapter provides adequate information about quantum mechanics, quantum computing, and quantum algorithms with their advantages. Based on these, we examine the use of quantum computing for circuit sizing in some specific applications.

6.1 Quantum Computing

Quantum mechanics refers to the physical description and behavior of particles at the subatomic level. At this level, classical physics cannot be applied thus, different mathematical expressions are required. While extremely hard to control, quantum mechanics potentially offers many transformative benefits.

Two main characteristics of quantum mechanics are **superposition** and **entanglement**. Superposition refers to the state of a quantum unit, which can have simultaneous values at the same time. For example, consider a coin as a quantum unit, it can be both heads and tails concurrently. Only when we take an observation on it, which is called **measurement** in quantum physics, the coin can then be collapsed into heads only, or tails only. In other words, the coin is brought from a microscopic environment (quantum level) to a macroscopic one (classical physics level). Superposition can hence be a means to concurrent simulation or parallel computation that can further enhance the executing speed of any algorithm. Conversely, entanglement is a phenomenon occurring among quantum units, which creates non-physical connections that withstand any physical distance.

Advanced Techniques for Optimal Sizing of Analog Integrated Circuits: Quantum Computing, Machine Learning, and Bio-inspired Optimization, First Edition.
Trang Hoang, Thinh Quang Do, Thang Quoc Nguyen, Hoang Trong Nguyen, Lihong Zhang, Octavia A. Dobre, and Trung Q. Duong.

6.1.1 Quantum States

Classical computing uses bits as the basic unit of information, which has a state of 0 or 1, the symbolic representations of the electric potential levels of electronic devices. On the other hand, quantum computing utilizes the quantum bit (qubit), which not only has the states $|0\rangle$ and $|1\rangle$ corresponding to the states of 0 and 1 for classical bits, but also has the states other than $|0\rangle$ or $|1\rangle$. These states, called the superposition states of a qubit, differentiate between qubits and classical bits [1].

The state of general qubit $|\psi\rangle$ can be any linear superposition state of $|0\rangle$ and $|1\rangle$ state, which can be expressed as

$$|\psi\rangle = \alpha\,|0\rangle + \beta\,|1\rangle\,, \tag{6.1}$$

where $|\alpha|^2 + |\beta|^2 = 1$, and $|\alpha|^2$ and $|\beta|^2$ are the probability that $|\psi\rangle$ collapses to $|0\rangle$ and $|1\rangle$, respectively.

Figure 6.1 further clarifies the representation of a single qubit. Notice that whenever a qubit is measured or observed, its superposition collapses, and that qubit then becomes a bit.

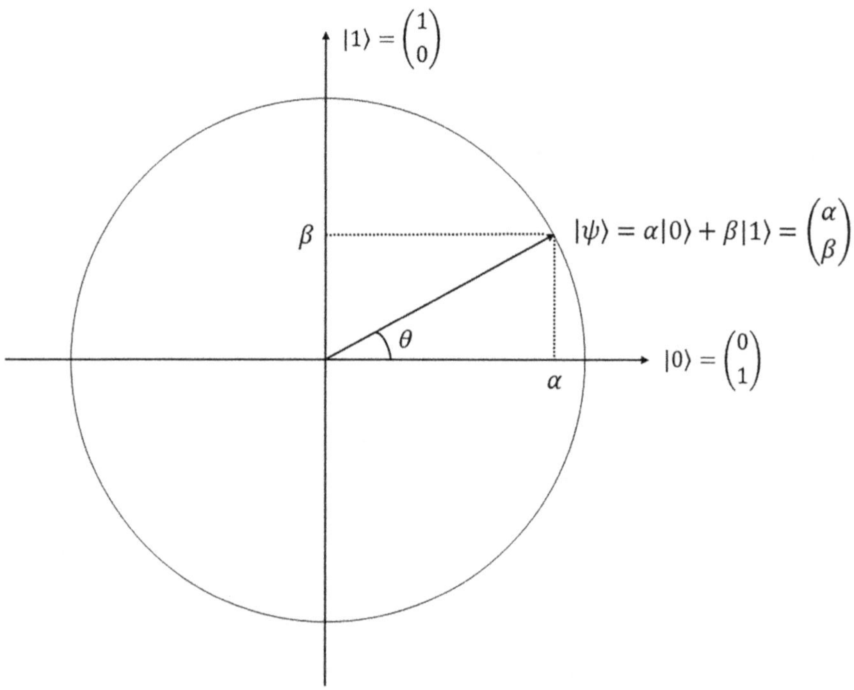

Figure 6.1 Representation of a single qubit.

Moreover, according to the Dirac notation, $|0\rangle$ and $|1\rangle$ can also be represented in the vector form by two-by-one complex matrices [2] as

$$|0\rangle = \begin{pmatrix} 1 \\ 0 \end{pmatrix}, |1\rangle = \begin{pmatrix} 0 \\ 1 \end{pmatrix}. \tag{6.2}$$

Therefore, $|\psi\rangle$ can be rewritten in the vector form of

$$|\psi\rangle = \alpha \begin{pmatrix} 1 \\ 0 \end{pmatrix} + \beta \begin{pmatrix} 0 \\ 1 \end{pmatrix} = \begin{pmatrix} \alpha \\ \beta \end{pmatrix}. \tag{6.3}$$

6.1.2 The Bloch Sphere

In the Bloch spherical coordinates, the qubit state $|\psi\rangle$ can be defined as

$$|\psi\rangle = \cos\left(\frac{\theta}{2}\right)|0\rangle + e^{i\phi}\sin\left(\frac{\theta}{2}\right)|1\rangle, \tag{6.4}$$

where $\cos\left(\frac{\theta}{2}\right)$ and $e^{i\phi}\sin\left(\frac{\theta}{2}\right)$ are complex numbers representing probability amplitudes of the qubit, θ and ϕ are the polar angle and azimuthal angle of a spherical coordinate, respectively.

In the Bloch spherical coordinates, two parameters, namely θ and ϕ, are required to describe a qubit [3]. Each real value of θ and ϕ corresponds to a three-dimensional point P of a particular qubit on the Bloch sphere [1], as illustrated in Figure 6.2.

6.1.3 Quantum Gates

A quantum gate is a gate that performs a unitary operation to manipulate or change a qubit. To obtain the new quantum state after applying such a transformation, the matrix representation of the quantum gate is multiplied by the initial qubit vector. There are various quantum gates used for quantum computing and quantum algorithms; however, this section presents fundamental quantum gates for quantum computing applications to analog circuit sizing optimization, namely Hadamard, rotation, and Pauli gates.

6.1.3.1 Hadamard Gate

Hadamard gate is a 1-qubit gate defined by the transformation matrix of

$$H = \frac{1}{\sqrt{2}} \begin{pmatrix} 1 & 1 \\ 1 & -1 \end{pmatrix}. \tag{6.5}$$

One of the most common applications of the Hadamard gate is to map n qubits initialized with $|0\rangle$ to a superposition of all 2^n orthogonal states in the standard $|0\rangle, |1\rangle$ basis with equal probability amplitudes. In other words, if being measured, the probability that the state collapses to either $|0\rangle$ or $|1\rangle$ is equal.

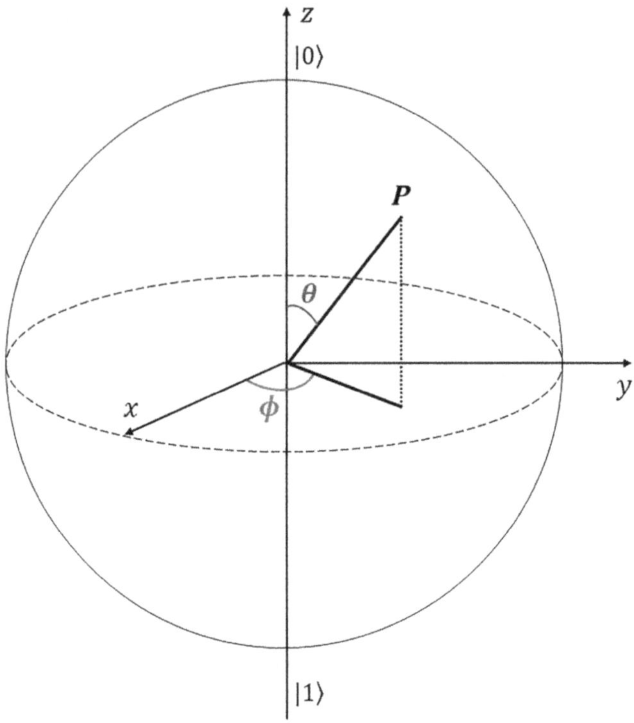

Figure 6.2 Representation of a single qubit on the Bloch sphere.

6.1.3.2 Rotation Gate

The rotation of a single qubit by an angle θ to change its state is defined by the transformation matrix of

$$U(\theta) = \begin{pmatrix} \cos(\theta) & -\sin(\theta) \\ \sin(\theta) & \cos(\theta) \end{pmatrix}. \tag{6.6}$$

There are three commonly used rotation gates, R_x, R_y, R_z, which rotate a qubit about the x-, y-, and z-axes of the Block sphere. Their operations are described by the following matrices

$$R_x(\theta) = \begin{bmatrix} \cos\left(\frac{\theta}{2}\right) & -i\sin\left(\frac{\theta}{2}\right) \\ -i\sin\left(\frac{\theta}{2}\right) & \cos\left(\frac{\theta}{2}\right) \end{bmatrix},$$

$$R_y(\theta) = \begin{bmatrix} \cos\left(\frac{\theta}{2}\right) & -\sin\left(\frac{\theta}{2}\right) \\ \sin\left(\frac{\theta}{2}\right) & \cos\left(\frac{\theta}{2}\right) \end{bmatrix},$$

$$R_z(\theta) = \begin{bmatrix} e^{-i\frac{\theta}{2}} & 0 \\ 0 & e^{i\frac{\theta}{2}} \end{bmatrix}.$$

6.1.3.3 Pauli Gates

Pauli gates refer to the set of single-qubit gates that perform a 180° rotation around one of the three basic coordinate axes in the Bloch sphere. These gates correspond to the Pauli matrices in linear algebra, which are

$$X = \begin{pmatrix} 0 & 1 \\ 1 & 0 \end{pmatrix}, Y = \begin{pmatrix} 0 & -i \\ i & 0 \end{pmatrix}, Z = \begin{pmatrix} 1 & 0 \\ 0 & -1 \end{pmatrix}. \tag{6.7}$$

The Pauli X gate swaps the probability of collapsing to $|0\rangle$ and $|1\rangle$, namely $|\alpha|^2$ and $|\beta|^2$. For the standard basis $|0\rangle$ and $|1\rangle$, the Pauli X gate performs the Boolean NOT operation that maps $|0\rangle \rightarrow |1\rangle, |1\rangle \rightarrow |0\rangle$. Pauli Y and Pauli Z gates do not have generic functions, but they both contribute to many practical quantum circuits.

6.1.4 Hadamard Product

The Hadamard product, also referred to as element-wise product or entry-wise product, is an operation that applies to two matrices of identical dimensions, resulting in a matrix where each element is the product of the corresponding elements from the input matrices [4].

Given two $m \times n$ matrices A and B, the Hadamard product $A \odot B$ results in a matrix of equal size $m \times n$, with elements calculated as

$$(A \odot B)_{ij} = (A)_{ij} \times (B)_{ij}, \tag{6.8}$$

where $i \in \{1, 2, \dots, m\}, j \in \{1, 2, \dots, n\}$ are the corresponding row and column index of the two matrices.

The Hadamard product is undefined in the case that the two matrices have different dimensions ($m \times n$ and $p \times q$, where $m \neq p$ or $n \neq q$).

6.1.5 Frobenius Norm

Considering an $m \times n$ matrix A, the Frobenius norm, or the Euclidean norm, is the matrix norm defined as the square root of the sum of the absolute squares of its elements. In other words, the Frobenius norm of an $m \times n$ matrix A is mathematically expressed as [5]

$$||A||_F = \sqrt{\sum_{i=1}^{m} \sum_{j=1}^{n} |a_{ij}|^2}, \tag{6.9}$$

where $i \in \{1, 2, \dots, m\}$ and $j \in \{1, 2, \dots, n\}$ are the corresponding row and column index, a_{ij} are the corresponding elements of the given matrix.

6.2 Quantum Variational Optimization

Variational algorithms typically begin by encoding classical data into quantum states via a feature map, a critical process that initializes the quantum circuits for subsequent computation. By exploiting quantum superposition and entanglement properties, these algorithms then apply a series of transformations aimed at optimizing the objective function. Upon measurement of the optimized quantum states, the resulting classical data provide critical insights for adjusting and refining the quantum circuit in subsequent iterations. A typical variational quantum optimizer consists of the following stages:

1) Initialize the system
2) Prepare an ansatz
3) Compute outcomes' probabilities (readout)
4) Evaluate cost function
5) Optimize variational parameters.

6.2.1 Initialize the System

In the initial stage, classical data are transformed into quantum states, which serve as reference states or the starting point for the optimization problem. If an educated guess for these reference states has been established from previous trials, starting the optimization from these states can lead to faster convergence of the variational quantum circuits. This is especially useful in quantum ML models, which often require numerous training iterations to achieve the desired results. If no training input data are available, beginning with a reference state and omitting the classical-to-quantum conversion step is preferable. In this chapter, three primary methods are discussed: amplitude embedding, angle encoding, and reference state construction.

6.2.1.1 Method 1. Amplitude Embedding

Typically, input data are characterized by multiple features, represented as $\mathbf{x} = \{x_1, x_2, \dots, x_M\}$, where each x_i denotes the value of a specific feature or component. This vector could represent anything from the load capacitor of a topology to features in a dataset and the width and length of a MOSFET in IC design. Therefore, different data sizes should be handled appropriately due to the variation of data dimension N. For quantum systems with n qubits, the data dimension N must be 2^n. If the original data does not meet this condition, preprocessing steps such as data padding or dimensionality reduction must be performed to align with the system's constraints. Once data is represented properly, it transforms into quantum states under the amplitude embedding method as follows:

$$|\psi\rangle = \sum_{j=1}^{N} \frac{x_j}{\sqrt{\sum_{u=1}^{N} |x_u^2|}} |j\rangle . \tag{6.10}$$

Here, $|j\rangle$ represents the computational basis states, and normalization ensures the quantum state has a unit norm, corresponding to a total probability of 1.

6.2.1.2 Method 2. Angle Encoding

This method maps classical data to quantum states using rotation gates, effectively bridging classical information with quantum computation. In this approach, the classical data serve as elements that contribute to constructing the rotation angles but are not the angles themselves. These angles are then applied to the standard quantum basis. There are two main steps in this encoding method. Firstly, the classical data $\mathbf{x}$ needs to be normalized, typically in the range $[0, 1)$, for feasible angle rotation. In the second step, different choices of rotation gates R_x, R_y, or R_z would be applied on standard basis states such as $|0\rangle$ depending on the specific encoding requirements, as follows:

$$\rho = U_R |0\rangle = Rot(\theta) |0\rangle , \tag{6.11}$$

where angle θ is built upon the input real-valued data x, represented by its binary format. For encoding, where it is necessary to change both the real and imaginary parts of the quantum state, the $R_x(\theta)$ gate becomes a primary operator. If we apply $R_x(2\pi x)$ (i.e., a $2\pi x$ rotation) to the standard basis state $|0\rangle$, the result will be

$$\rho = R_x(2\pi x) |0\rangle = \begin{bmatrix} \cos(\pi x) & -i\sin(\pi x) \\ -i\sin(\pi x) & \cos(\pi x) \end{bmatrix} \begin{bmatrix} 1 \\ 0 \end{bmatrix} = \cos(\pi x) |0\rangle - i\sin(\pi x) |1\rangle . \tag{6.12}$$

Rotating the standard basis state $|0\rangle$ by a $2\pi x$ angle allows real-valued data to be directly encoded into quantum states, enabling the control of the qubit's superposition. For data that primarily affects probability amplitudes, such as binary or normalized data where the goal is to distribute the probability between $|0\rangle$ and $|1\rangle$, the $R_y(\theta)$ gate is typically the best choice. The R_y gate rotates the qubit around the y-axis of the Bloch sphere and controls the qubit's probability amplitudes without introducing a phase factor, unlike the R_x gate.

$$\rho = R_y(2\pi x) |0\rangle = \begin{bmatrix} \cos(\pi x) & -\sin(\pi x) \\ \sin(\pi x) & \cos(\pi x) \end{bmatrix} \begin{bmatrix} 1 \\ 0 \end{bmatrix} = \cos \pi x |0\rangle + sin\pi x |1\rangle . \tag{6.13}$$

It is less common to use R_z gate for angle encoding since it introduces a phase factor but does not change the relative amplitudes of $|0\rangle$ and $|1\rangle$. This can be useful in specific quantum algorithms where manipulating phase information is the key to the algorithm's success. For instance, applying the $R_z(2\pi x)$ gate to the basis standard state $|0\rangle$ does not affect the probability of measure $|0\rangle$ but its phase information, by an angle proportion to x

$$\rho = R_z(2\pi x)\,|0\rangle = \begin{bmatrix} e^{-i\pi x} & 0 \\ 0 & e^{-i\pi x} \end{bmatrix} \begin{bmatrix} 1 \\ 0 \end{bmatrix} = e^{-i\pi x}\,|0\rangle\,. \tag{6.14}$$

6.2.1.3 Method 3. Reference State Construction

In methods 1 and 2, classical data undergoes a specific conversion process before being manipulated by the variational circuit. However, in some scenarios-such as unsupervised learning-there may be no predefined input dataset. In these cases, method 3 becomes applicable, where the process begins with a reference quantum state rather than encoded classical data. This allows the system to evolve and explore patterns without the need for external data input.

A reference state ρ is defined as $\rho = U_R\,|0\rangle$ where U_R is a non-parameterized unitary operator [6]. While there is no fixed choice for the reference state, selecting it wisely can significantly accelerate the learning process in a quantum model. Normally, the default state $|0\rangle^{\otimes n}$ is employed in a n-qubit quantum circuit due to its simplicity. In this case, $U_R \equiv I$. However, if one wishes to begin with a superposition state, such as $\frac{1}{\sqrt{2}}(|00\rangle + |11\rangle)$ in a 2-qubit system, a different approach is needed. One approach is to apply a Hadamard gate to qubit 0, creating a superposition state (H_0). Following this, a CNOT gate with control qubit 0 regulates the target qubit 1 ($CNOT_{01}$), which is used to entangle the two qubits. This sequence of operations creates the desired entangled superposition state. Therefore, the reference state can be obtained by applying the unitary operator $U_R = CNOT_{01}H_0$, to the initial state $|00\rangle$, resulting in the superposition state

$$U_R = CNOT_{01}H_0\,|00\rangle = \frac{1}{\sqrt{2}}(|00\rangle + |11\rangle). \tag{6.15}$$

6.2.2 Prepare an Ansatz

The key feature of a variational quantum circuit lies in its focus on exploring variations among different quantum states. Constructing a variational circuit facilitates the systematic analysis of these state differences by mapping them through a set of tunable parameters. This flexibility is essential for optimization tasks, where the goal is to find an optimal state by adjusting the parameters to minimize or maximize a cost function.

6.2.2.1 Parameterized Quantum Operator

A parameterized quantum circuit is constructed using quantum gates that are controlled by a set of m tunable parameters, denoted as $\theta = \theta_0, \theta_1, \ldots, \theta_{m-1}$.

The primary objective of the parameterized quantum circuit is to guide an initial reference state ρ, prepared during the initialization phase described in Section 6.2.1, toward a desired target state $\psi(\theta)$, as follows:

$$\begin{aligned}\psi(\theta) &= U_V(\theta)\rho = U_V(\theta)U_R\,|0\rangle \\ &= U_A(\theta)\,|0\rangle\,,\end{aligned} \tag{6.16}$$

where U_V represents the variational operator, $U_A(\theta) = U_V(\theta)U_R$ is known as an ansatz. Notably, in certain cases, such as the amplitude embedding method, there is no U_R, then $U_A \equiv U_V$. By fine-tuning the set of k parameters θ, there is no need for the initialization operator U_R, resulting in $U_A \equiv U_V$.

In an IC sizing optimization problem, where n qubits are employed to represent the quantum information of the widths, lengths, and other information of CMOS transistors in a circuit, the transition between classical and quantum representations typically requires the use of a *density matrix*. This approach increases the total dimensionality of the system to 2^{2n}, as the density matrix consists of all possible quantum states in a 2^n-dimensional Hilbert space. As n grows larger, the complexity of the problem expands exponentially, a phenomenon commonly referred to as the *curse of dimensionality*. To mitigate this challenge, it becomes essential to impose constraints, allowing for the construction of a *truncated ansatz* that restricts the search space and reduces the computational load. One of the most popular methods is the heuristic ansatz, which can be applied to general IC design problems.

6.2.2.2 Heuristic Ansatze and Trade-Offs

When specific information about the IC design circuit is unavailable to help manage the dimensionality, an arbitrary family of parameterized circuits with fewer than 2^{2n} parameters can be utilized. However, reducing the parameter space carries the risk of excluding the true solution to the problem, potentially resulting in suboptimal solutions. This trade-off between the time required to find the solution and the quality or precision of the results should be carefully considered [6]. Fortunately, in analog optimization problems, many approximation models have been developed to estimate practical scenarios; therefore, this trade-off is likely to be minor compared to the overall benefits.

6.2.2.3 N-Local Circuits

One of the widely used families of heuristic ansatze is the N-local circuits [6], where N denotes the maximum size of the gates used within the entire circuit and N is lower than the number of utilized qubits. The N-local circuits are characterized by their two-layered structure: the rotation layer and the entanglement layer. In the rotation layer, standard rotation gates such as R_x, R_y, and R_z are stacked on top of each other. In some variations of N-local circuits, multiple rotation layers

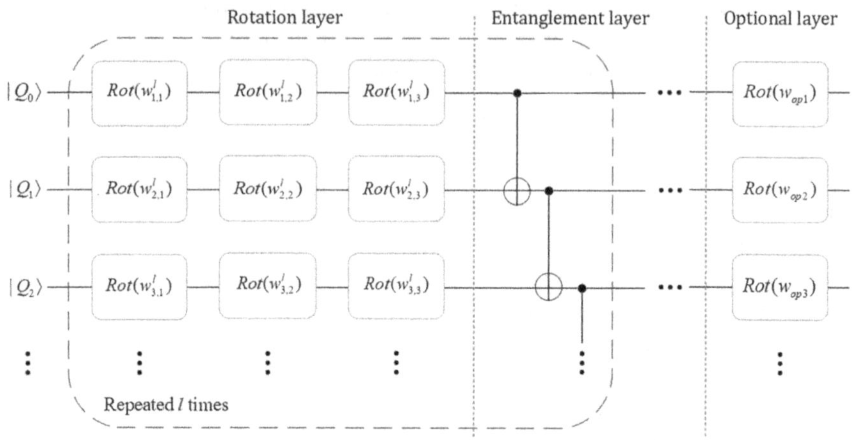

Figure 6.3 A typical N-local circuit with l layers.

are used in series to enhance the flexibility and capacity of the circuit. The entanglement layer, on the other hand, includes entanglement gates such as CNOT and Toffoli, which are employed according to an entanglement strategy to capture and manage crucial correlations between the qubits. Additionally, an optional rotation layer can be added at the end of the circuit to provide more flexibility.

In N-local circuits, both the rotation and entanglement layers can be either parameterized or non-parameterized. However, it is crucial that at least one of these layers is parameterized to introduce variability into the circuit, which is a key aspect of variational quantum circuits. Figure 6.3 describes a 2-local circuit with single-qubit rotation gates and 2-qubit entanglement gates.

In this 2-local circuit, the rotation layer consists of three sublayers, each applying rotations to individual qubits with varying angles, allowing for greater flexibility in the optimization process. The choice of rotation gates is influenced by both the optimization objectives and the properties of the gates themselves (as outlined in Section 6.2.1.2). The number of gates applied to each qubit may vary and is typically determined through a trial-and-error approach. However, it is crucial to avoid an excessive number of gates, as this can impede convergence and reduce the overall efficiency of the circuit.

6.2.3 Readout

The primary goal of variational quantum algorithms is not to obtain a single collapsed outcome through measurement but to understand the overall behavior of the quantum system by examining the probability distribution of possible measurement outcomes. A single measurement provides limited information, whereas understanding the full distribution helps guide the optimization process and find optimal solutions probabilistically, improving the algorithm's performance. To facilitate this analysis, an observable $\hat{H} = \hat{H}^\dagger$ is applied, projecting the optimal state $|\psi(\theta)\rangle$ from Section 6.2.2 onto one of the eigenstates $|\lambda\rangle$ of $\hat{H}$. This measurement yields the corresponding eigenvalue λ as the outcome, known as the *readout*.

6.2.3.1 Method 1. Sampler

The sampler method executes the quantum circuit that prepares the optimal state $|\psi(\theta)\rangle$ multiple times on a quantum device. The cumulative measurement results provide an estimation of the probability of obtaining any basis state $|b\rangle$, reflecting the likelihood that the state will collapse into a classical bit string b that represents the outcome of the overall run, as follows:

$$p_b = |\langle b|\psi(\theta)\rangle|^2, \tag{6.17}$$

where b can be any integer binary string $b \in \{0, 1, \ldots, 2^n - 1\}$, given that $N = 2^n$ represents all possible states of $|\psi(\theta)\rangle$. This classical information b can be decoded into the value of resistors, capacitors, widths, and lengths of MOSFETs, where any configurations of the IC topology or a function contain all of them. However, the primary focus is on the probability p_b that $|\psi(\theta)\rangle$ will collapse into these values, rather than on the specific outcome b itself.

6.2.3.2 Method 2. Estimator

The estimator method calculates the expectation value, which is the average of all possible outcomes λ of a measurement, weighted by their likelihood. This is expressed as

$$\hat{H}_\psi = \langle\psi(\theta)|\,\hat{H}\,|\psi(\theta)\rangle = \sum_\lambda p_\lambda \lambda, \tag{6.18}$$

where $\hat{H}_\psi$ represents the expectation value of the observable $\hat{H}$ and the rightmost expression sums over all possible outcomes, each multiplied by its corresponding probability $p_\lambda = |\langle\lambda|\psi(\theta)\rangle|^2$. By associating the observable $\hat{H}$ with design performance such as power consumption, maximum chip frequency, or available bandwidth, the expectation value serves as a measure of how well a given IC topology meets optimization goals. Additionally, this value can be directly integrated into a cost function to evaluate various configurations of the IC, guiding the design process toward optimal solutions. However, the selection of $\hat{H}$ can become extremely complex, particularly when dealing with strict specifications. The need to integrate multiple design objectives within a single observable can lead to challenges in formulation, requiring careful consideration of the interactions between different performance metrics. Consequently, defining $\hat{H}$ becomes a critical step that necessitates a thorough understanding of both the quantum system and the classical design parameters.

6.2.4 Evaluate Cost Function

In IC design, customers often provide detailed specifications regarding performance metrics such as gain, bandwidth, and power consumption. These requirements serve as the guiding benchmarks for the IC designers, who adjust the configurations of the IC topology to meet the desired performance. The role of the cost function in this optimization process is to quantify how closely a given IC design configuration satisfies these specifications, guiding the variational quantum algorithm to search for optimal solutions.

Regarding the sampler method, the classical bit strings b represent various design configurations (e.g., MOSFET widths, lengths, or other parameters). Each bit string b can be decoded into a set of physical performance metrics such as power consumption or gain. Therefore, the cost function should compare the physical performance resulting from the sampled configurations to the specification. Let $F(b_i)$ be the function that maps a bit string b_i to a corresponding performance metric (e.g., voltage gain) and F^* be the desired performance metric. The cost function, which is computed using a modified weighted mean square approach where p_{b_i} serves as the weight of $F(b_i)$, can be expressed as

$$C(\theta) = \sum_i p_b(F(b_i) - F^*)^2. \tag{6.19}$$

This probabilistic weighting effectively directs the variational quantum algorithm toward configurations that are not only likely but also optimal in terms of performance metrics, thus contributing to the convergence of the algorithm to an optimal solution, ensuring that the design goals are met efficiently.

6.2.5 Optimize Variational Parameters

Once the cost function has been evaluated, it serves as feedback for the classical optimization process. Subsequently, classical optimizers are employed to search across the entire domain of variational parameters θ, aiming to find the global minimum of $C(\theta)$. The ultimate objective of the optimization is to reduce the cost function as close to zero as possible, i.e., $C(\theta) \approx 0$, which means that the IC design configuration is in near-optimal alignment with the given specifications. One common approach in optimizing the cost function is to employ stochastic gradient descent (SGD), provided that the gradient of the function $\nabla_\theta C(\theta)$ is accessible. This method updates the parameters $\vec{\theta}$ iteratively according to the following rule

$$\theta_{n+1} = \theta_n - \eta \nabla_\theta C(\theta_n), \tag{6.20}$$

where $\nabla_{\vec{\theta}} C(\vec{\theta}_n)$ represents the gradient of the cost function with respect to the parameters $\vec{\theta}$ at iteration n, and η is the learning rate. In some cases, the goal is to optimize the target function rather than minimize the cost function. In these instances, SGD can still be effective. If the target function is denoted by $T(\vec{\theta})$, we can obtain a near-optimal solution by finding the global minimum of $R(\vec{\theta}) = -T(\vec{\theta})$.

6.3 Grover Optimizer

A significant advantage that sets quantum computing apart from classical computing is its ability to simultaneously access and process multiple elements at unparalleled speeds. A prime example is Grover's algorithm, a prominent searching algorithm in unstructured databases with N elements, which offers a quadratic speedup over conventional methods. The complexity of a search task is reduced from $O(N)$ as in classical approaches to $O(\sqrt{N})$ with Grover's algorithm. As the size of the database grows, this method's efficiency amplifies accordingly. The underlying mechanism of Grover's algorithm is the amplitude amplification trick, a technique also utilized by other quantum algorithms such as Deutsch–Jozsa and Bernstein–Vazirani. This technique allows Grover's algorithm to systematically amplify the probability of finding the correct item, thereby achieving its remarkable speedup.

The superiority in search speed becomes clearer through a detailed analysis of the following search problem, represented by the abstract function $f(x)$, which queries the database for the item x. In this model, if the item x is the key sought in the search task, then $f(x) = 1$, indicating a successful match. Conversely, if x is not the key, then $f(x) = 0$. The black-box search problem can be alternatively expressed as $f(x) : \{0, 1\}^n \rightarrow \{0, 1\}$, where n is the bit-size of the database. The goal is to identify a solution x_0 within the database such that $f(x_0) = 1$. The number of times the function $f(x)$ must be queried serves as an indicator of the algorithm's complexity.

In classical methods, the worst-case scenario unfolds when every single one of the $N - 1$ items (with $N = 2^n$) must be queried without uncovering the sought item. In this case, the final query will ultimately yield the solution x_0. In contrast, Grover's algorithm, with its quadratic speedup, significantly reduces this number. Instead of requiring N queries, Grover's algorithm achieves the same result with $\sqrt{N}$ queries. This substantial reduction in the number of queries highlights the remarkable efficiency of quantum search algorithms, making them particularly valuable for tackling large-scale search problems where classical methods become increasingly impractical.

Grover's algorithm can be effectively employed in analog IC design optimization to significantly enhance the efficiency of the search for optimal circuit configurations. In analog IC design, finding the optimal configuration that meets the desired specifications is often equivalent to a search problem across a complex solution space of multiple design parameters, such as transistor widths and lengths, resistors, and capacitors. While the variational quantum algorithm focuses on optimizing the cost function by fine-tuning circuit parameters to achieve an optimal set of features for a given IC topology, Grover's algorithm uses a different approach called amplitude amplification. Grover's approach systematically amplifies the probability of identifying the optimal set, thus making the search process considerably faster and more efficient, instead of exhaustively simulating all possible combinations of parameter values.

Grover's algorithm can be divided into the following steps:

1) Prepare the initial quantum state.
2) Apply the Oracle operator.
3) Implement the Grover diffusion operator.
4) Repeat steps 2 and 3 for $m_{optimal}$ iterations.

6.3.1 Initialization

The searching process operates on 2^n distinct inputs; thus, it requires n data qubits along with a target qubit, all of which are initialized at the state $|0\rangle$. The role of this target qubit will be comprehensively discussed in Section 6.3.2, where it plays a crucial part in enabling the phase kickback mechanism. Then, Hadamard gates are applied on n input qubits to achieve uniform superposition of 2^n states as follows:

$$\begin{aligned}|\psi_{data}\rangle = |\psi_{eq}\rangle = H^{\otimes n}|0\rangle^{\otimes n} &= \frac{1}{\sqrt{2^n}}(|00\ldots0\rangle + |00\ldots1\rangle + \ldots + |11\ldots1\rangle) \\ &= \frac{1}{\sqrt{2^n}}\sum_{x=0}^{N-1}|x\rangle . \end{aligned} \tag{6.21}$$

This condition provides a fair starting point, free from bias toward any potential outcome. The target qubit is flipped under the quantum NOT gate operation before undergoing the Hadamard transformation:

$$HX|0\rangle = H|1\rangle = \frac{1}{\sqrt{2}}(|0\rangle - |1\rangle). \tag{6.22}$$

This operation appropriately prepares the target qubit for later procedures. Afterward, combining these two components of Eqs. (6.21) and (6.22), the resulting state of the system is expressed as

$$|\psi_1\rangle = \frac{1}{\sqrt{2^n}}\sum_{x=0}^{N-1}|x\rangle \otimes \frac{1}{\sqrt{2}}(|0\rangle - |1\rangle). \tag{6.23}$$

6.3.2 Oracle Operator

Quantum oracle operator, denoted as U_f, is formulated such that when applied to the input state $|x\rangle \otimes |y\rangle$, it yields the output state $U_f |x\rangle |y\rangle = |x\rangle \otimes |f(x) \oplus y\rangle$, where the function $f(x)$ returns 1 for the winner item (sought item) and 0 for all the others. In the context of analog IC design, this function $f(x)$ can be mapped to a variety of performance metrics, such as gain, bandwidth, or power consumption. For example, $f(x) = 1$ would indicate that a particular configuration of design parameters (e.g., transistor widths and lengths) satisfies the desired performance specifications, while $f(x) = 0$ would mean the configuration does not meet the design goals. This formulation makes the Oracle operator central to using Grover's algorithm for optimization in analog IC design, as it directly encodes the feasibility of different parameter sets with respect to the given performance criteria. Applying this operator U_f on the state $|x\rangle \otimes (|0\rangle - |1\rangle)$ results in the following state:

$$\begin{aligned} U_f |x\rangle (|0\rangle - |1\rangle) &= |x\rangle \otimes (|f(x) \oplus 0\rangle - |f(x) \oplus 1\rangle) \\ &= |x\rangle \otimes (|f(x)\rangle - |\overline{f(x)}\rangle) \\ &= (-1)^{f(x)} |x\rangle \otimes (|0\rangle - |1\rangle). \end{aligned} \tag{6.24}$$

It is crucial to note that the oracle operation allows the appearance of $f(x)$ in the global phase of the state, which can be referred to as *phase kickback trick* [7]. Undergoing the transformation of U_f, the obtained state $|\psi_1\rangle$ from the initialization step transitions to

$$|\psi_2\rangle = U_f |\psi_1\rangle = \frac{1}{\sqrt{2^n}} \sum_{x=0}^{2^n-1} (-1)^{f(x)} |x\rangle \otimes \frac{1}{\sqrt{2}} (|0\rangle - |1\rangle). \tag{6.25}$$

It is evident that the target qubit remains unchanged under the action of the oracle operator. Therefore, the target qubit can be omitted, and from this point onward, the Oracle operation is represented as $U_f |x\rangle = (-1)^{f(x)} |x\rangle$, indicating that it exclusively acts on the computational basis states, leaving the target qubit unchanged.

The uniform superposition state $|\psi_{data}\rangle$ can be conceptualized as existing within a vector space spanned by two mutually orthogonal vectors: the searched (winner) item $|w\rangle$ and other remaining (loser) items $|l\rangle$ (Figure 6.4a). These vectors are mutually orthogonal because they represent distinct and non-overlapping states in the Hilbert space: $|w\rangle$ identifies the unique solution marked by the oracle, while $|l\rangle$ corresponds to all other non-solution states. This orthogonality can be written as $|w\rangle \perp |l\rangle$ or, equivalently, $|l\rangle = |w^{\perp}\rangle$, emphasizing that the subspace of loser states is entirely orthogonal to the subspace containing the winner state. To elaborate further on this concept, the items $|l\rangle$ can be obtained by excluding the winner

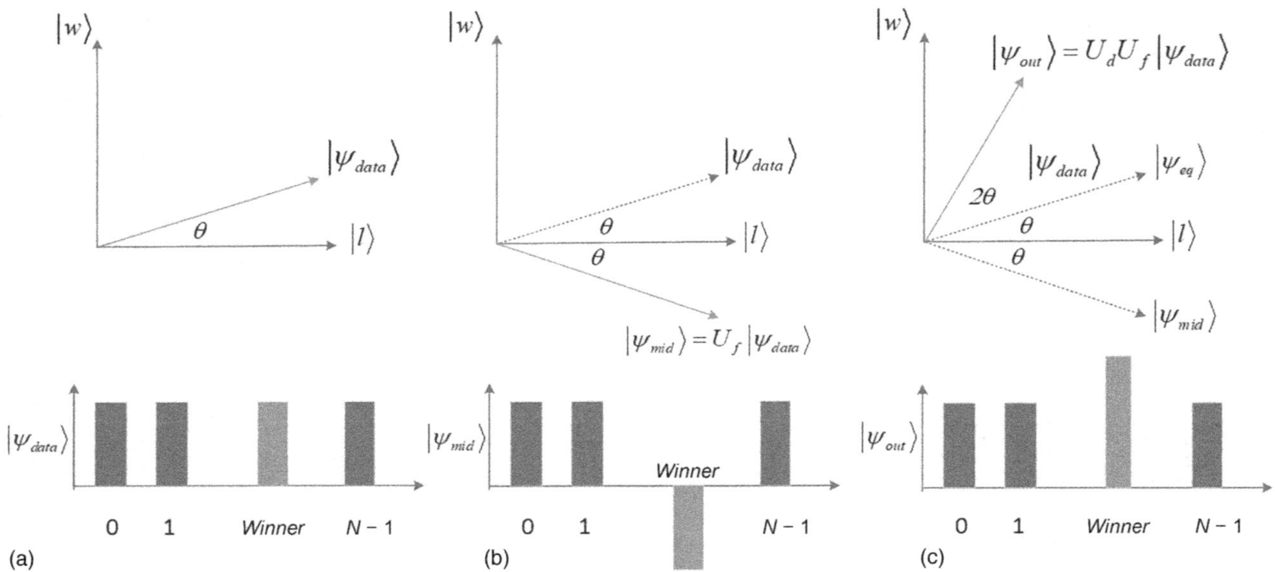

Figure 6.4 Geometric representation of Grover's algorithm. (a) Initial amplitudes, (b) amplitudes after U_f transform, and (c) amplitudes after $U_d U_f$ transform. Source: Pattanayak [7]/with permission of Springer Nature.

item $|w\rangle$ from the superposition state $|\psi_{data}\rangle$. Therefore, ψ_{data} can be alternatively represented as

$$|\psi_{data}\rangle = \frac{1}{\sqrt{2^n}} |w\rangle + \frac{\sqrt{2^n - 1}}{\sqrt{2^n}} |l\rangle . \tag{6.26}$$

Among the 2^n elements of the database, there is exactly one winner element, resulting in the corresponding coefficient $\langle w|\psi_{data}\rangle = \frac{1}{\sqrt{2^n}}$. In other contexts, there may be multiple winners, leading to the coefficient $\langle w|\psi_{data}\rangle = \frac{\alpha}{\sqrt{2^n}}$, where α represents the number of winner elements. For the purposes of a fundamental analysis, this book considers the case of a single marked element. Utilizing trigonometric principles, the data state can be expressed as

$$|\psi_{data}\rangle = \sin\theta\, |w\rangle + \cos\theta\, |l\rangle , \tag{6.27}$$

where $\sin\theta = \frac{1}{\sqrt{2^n}}$ and $\cos\theta = \frac{\sqrt{2^n-1}}{\sqrt{2^n}}$. This representation ensures that the total probability condition is satisfied, as $\sin^2\theta + \cos^2\theta = \frac{1}{2^n} + \frac{2^n-1}{2^n} = 1$. The phase kick back operation of U_f introduces a negative phase to the winner action $|w\rangle$ while preserving the phase associated with the remaining actions $|l\rangle$, and hence the output state $|\psi_{mid}\rangle$ can be expressed as

$$|\psi_{mid}\rangle = -\sin\theta\, |w\rangle + \cos\theta\, |l\rangle . \tag{6.28}$$

Figure 6.4b shows the operation of U_f on $|\psi_{data}\rangle$. According to [8], the operator U_f can be written as

$$U_f = I - 2(|w\rangle\langle w|). \tag{6.29}$$

A critical challenge in applying Grover's algorithm for analog IC optimization is the construction of the oracle operator U_f. This process requires a full simulation for each configuration in the search space to determine whether it is considered "good." When constructing the Oracle is as computationally expensive as evaluating all configurations through simulation—common in many analog IC design problems—Grover's algorithm may prove inadequate. Furthermore, the lack of accurate approximate models for manually evaluating the performance of IC topologies increases the overall cost of the process. This challenge opens an opportunity for solutions through ML. By training models to predict whether a given circuit configuration meets design requirements, a more efficient approach to constructing the oracle could be developed. These models could serve as faster alternatives to costly simulations, reducing computational overhead and enabling the effective use of Grover's algorithm in optimizing complex analog designs.

6.3.3 Grover Diffusion Operator

Grover iteration operator is the transformation that flips the vector $|\psi_{mid}\rangle$ about the vector $|\psi_{eq}\rangle$ where $|\psi_{eq}\rangle = H^{\otimes n}|0\rangle^n$ represents the equal superposition states (as demonstrated in Figure 6.4c). This makes Grover's operation become $U_d = (2|\psi_{eq}\rangle\langle\psi_{eq}| - I)$. According to [7], the diffusion operator U_d reflecting about the vector $|\psi_{data}\rangle$ can be built upon Hadamard gates as

$$\begin{aligned} U_d &= H^{\otimes n}(2|0\rangle^{\otimes n}\langle 0|^{\otimes n} - I)H^{\otimes n} \\ &= 2(H^{\otimes n}|0\rangle^{\otimes n})(\langle 0|^{\otimes n}H^{\otimes n}) - H^{\otimes n}IH^{\otimes n} \\ &= 2|\psi_{eq}\rangle\langle\psi_{eq}| - I, \end{aligned} \tag{6.30}$$

where $(H^2)^{\otimes n} \equiv I$. For the practical implementation of U_d, the operation $(2|0\rangle^{\otimes n}\langle 0|^{\otimes n} - I)$ must be constructed using standard quantum gates. This can be accomplished by analyzing its unique effect on $|0\rangle^{\otimes n}$ and the other remaining basis states.

(a) When $|x\rangle = |0\rangle^{\otimes n}$

$$\begin{aligned} (2|0\rangle^{\otimes n}\langle 0|^{\otimes n} - I)\,|x\rangle &= (2|0\rangle^{\otimes n}\langle 0|^{\otimes n} - I)|0\rangle^{\otimes n} \\ &= 2|0\rangle^{\otimes n}\langle 0|0\rangle^{\otimes n} - |0\rangle^{\otimes n} \\ &= |0\rangle^{\otimes n} \\ &= |x\rangle\,. \end{aligned} \tag{6.31}$$

(b) When $|x\rangle \neq |0\rangle^{\otimes n}$

$$\begin{aligned} (2|0\rangle^{\otimes n}\langle 0|^{\otimes n} - I)\,|x\rangle &= 2|0\rangle^{\otimes n}\langle 0|^{\otimes n}\,|x\rangle - |x\rangle \\ &= 2|0\rangle^{\otimes n}.0 - |x\rangle \\ &= -\,|x\rangle\,. \end{aligned} \tag{6.32}$$

From (a) and (b), the operation of $(2|0\rangle^{\otimes n}\langle 0|^{\otimes n} - I)$ can be synthesized as

$$(2|0\rangle^{\otimes n}\langle 0|^{\otimes n} - I)\,|x\rangle = \begin{cases} |x\rangle\,, & |x\rangle = |0\rangle^{\otimes n}, \\ -\,|x\rangle\,, & |x\rangle \neq |0\rangle^{\otimes n}. \end{cases} \tag{6.33}$$

Equation (6.33) points out that $(2|0\rangle^{\otimes n}\langle 0|^{\otimes n} - I)$ is a conditional phase flip operator, which can be built upon different combinations of standard quantum gates depending on the value of employed qubits, n.

To gain a comprehensive understanding of U_d operation in the vector space spanned by $|w\rangle$ and $|l\rangle$, its geometric representation is depicted as

$$\begin{aligned} U_d &= 2|\psi_{eq}\rangle\langle\psi_{eq}| - I \\ &= 2\begin{bmatrix}\cos\theta \\ \sin\theta\end{bmatrix}\begin{bmatrix}\cos\theta & \sin\theta\end{bmatrix} - \begin{bmatrix}1 & 0 \\ 0 & 1\end{bmatrix} \end{aligned}$$

$$
\begin{aligned}
&= 2\begin{bmatrix} \cos^2\theta & \cos\theta\sin\theta \\ \sin\theta\cos\theta & \sin^2\theta \end{bmatrix} - \begin{bmatrix} 1 & 0 \\ 0 & 1 \end{bmatrix} \\
&= \begin{bmatrix} \cos^2\theta - \sin^2\theta & 2\cos\theta\sin\theta \\ 2\sin\theta\cos\theta & \sin^2\theta - \cos^2\theta \end{bmatrix} \\
&= \begin{bmatrix} \cos 2\theta & \sin 2\theta \\ \sin 2\theta & -cos2\theta \end{bmatrix}.
\end{aligned} \tag{6.34}
$$

Applying U_d on the state $\left|\psi_{mid}\right\rangle$, the system's output state transforms into

$$
\begin{aligned}
\left|\psi_{out}\right\rangle = U_d \left|\psi_{mid}\right\rangle &= \begin{bmatrix} \cos 2\theta & \sin 2\theta \\ \sin 2\theta & -\cos 2\theta \end{bmatrix} \begin{bmatrix} \cos\theta \\ -\sin\theta \end{bmatrix} \\
&= \begin{bmatrix} \cos 2\theta\cos\theta - \sin 2\theta\sin\theta \\ \sin 2\theta\cos\theta + \cos 2\theta\sin\theta \end{bmatrix} \\
&= \frac{1}{2}\begin{bmatrix} (\cos 3\theta + \cos\theta) + (\cos 3\theta - \cos\theta) \\ (\sin 3\theta + \sin\theta) + (\sin 3\theta - \sin\theta) \end{bmatrix} \\
&= \begin{bmatrix} \cos 3\theta \\ \sin 3\theta \end{bmatrix} \\
&= \sin 3\theta\,|w\rangle + \cos 3\theta\,|l\rangle\,.
\end{aligned} \tag{6.35}
$$

The output state reveals a notable increase in the magnitude of the winning state compared to the initial state $\left|\psi_{data}\right\rangle$ while that of other loser states decreases. This phenomenon can be attributed to the ascent of the sine function and the descent of the cosine function in the first quadrant.

6.3.4 Grover Iteration

Afterward, each Grover iteration G can be achieved by combining the two operators U_f and U_d in the subsequent steps 2 and 3, as follows:

$$
U_G = U_d U_f. \tag{6.36}
$$

By applying Grover operator successively for m iterations, the initial state ψ_{data} converges into the winning item $|w\rangle$, or $U_G^m\left|\psi_{data}\right\rangle = (U_d U_f)^m \left|\psi_{data}\right\rangle \approx |w\rangle$. According to [1], the optimal number of iterations m_{opt} can be calculated as

$$
m_{opt} = \left[\frac{\pi}{4}\sqrt{N} - \frac{1}{2}\right] = \left[\frac{\pi}{4}2^{\frac{n}{2}} - \frac{1}{2}\right], \tag{6.37}
$$

which reveals the core advantage of Grover's algorithm: instead of needing N queries to search through an unsorted database of size N, Grover's algorithm only requires approximately $\sqrt{N}$ iterations to locate the target item with high probability. This results in a significant reduction in the number of queries, especially

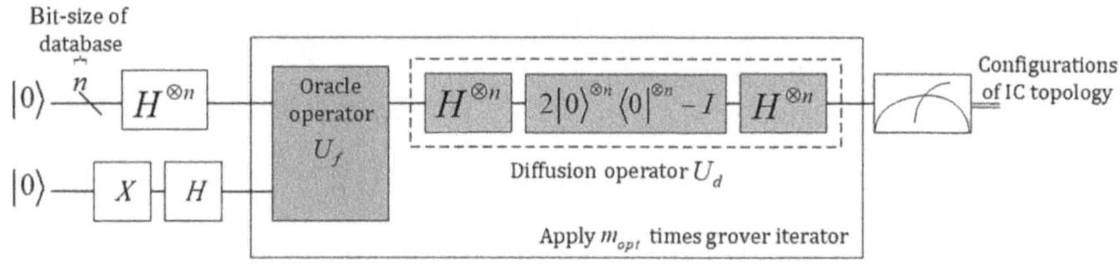

Figure 6.5 General circuit design of Grover's algorithm for searching an *n*-bit database. Source: Pattanayak [7]/with permission of Springer Nature.

for large N, where the difference between N and $\sqrt{N}$ becomes increasingly substantial. Figure 6.5 demonstrates the overall implementation of the Grover algorithm for n bit-size database.

The exponential speedup provided by Grover's algorithm 6.1 in finding the optimal solution among a vast database makes it an appealing candidate for optimization problems in IC analog design, particularly when searching for the best configuration of transistor parameters like widths and lengths. However, one of the major challenges lies in the construction of the oracle operator, which traditionally requires evaluating the performance of each parameter set to determine whether it satisfies the design specifications. This process can be computationally expensive, as it often involves simulating the analog circuit's behavior for each configuration, making it difficult to scale Grover's algorithm for large parameter spaces. Addressing this challenge involves developing a more efficient method for constructing the oracle, one that does not rely on exhaustive evaluations of the circuit's performance. Potential approaches include leveraging ML models or heuristics to predict whether a given configuration will meet the

Algorithm 6.1: Grover Algorithm

1 **Input:** Computational basis state $|0\rangle^{\otimes n}$, Target qubit $|0\rangle$, Identity matrix I, Database with $D = 2^n$

2 **Output:** Target element $|w\rangle$ with a high probability of successful measurement

3 Initialize $|\psi_1\rangle = H^{\otimes n}|0\rangle^{\otimes n} \otimes HX|0\rangle$

4 Repeat Grover iterations through Oracle and Diffusion Operator over m_{opt} times:

5 $\quad U_G|\psi_1\rangle \rightarrow \frac{1}{\sqrt{2^n}} \sum_{x=0}^{2^n-1} [(-1)^{f(x)}(2|\psi_{eq}\rangle\langle\psi_{eq}| - I)]^{m_{opt}}|x\rangle$

design goals, thus reducing the number of costly simulations required. Once these obstacles are overcome, this could lead to more efficient designs, faster time-to-market, and reduced development costs.

6.4 Quantum Annealing

Another potential quantum technique for optimization is quantum annealing (QA), which offers a heuristic quantum optimization algorithm to find the ground state of Ising models. Compared to classical optimization algorithms (such as GA, PSO, and DE), the QA algorithm, which has been recently tested in solving the optimization problem in semiconductor manufacturing, can avoid being trapped in local minima and has the potential to find better global solutions due to the tunnelling and superposition nature of qubits. Moreover, the QA algorithm is more robust to noise and other sources of errors than gate-based quantum algorithms [9].

6.4.1 Adiabatic Quantum Computing

In Sections 6.1–6.3, we have mainly focused on gate-based quantum computing, in which a quantum algorithm is implemented by combining several steps of discrete operations (or quantum gates). However, adiabatic quantum computing relies on the use of continuous transformations. As such, we will use a Hamiltonian $H(t)$ varying with time that will be the driving force to change the state of our qubits according to the time-dependent Schrödinger equation, as follows:

$$H(t)\,|\psi(t)\rangle = i\hbar\frac{\partial}{\partial T}\,|\psi(t)\rangle\,, \tag{6.38}$$

where $H(t)$ is the time-dependent Hamiltonian that represents the energy of the system, $|\psi(t)\rangle$ is the state vector of the system, i is the imaginary unit such that $i^2 = -1$, and $\hbar$ is the reduced Planck's constant.

In addition to using the time-dependent Hamiltonian, there is another element for our new quantum computing model: the idea of adiabatic evolution. Roughly speaking, an adiabatic process is one in which the "energy configuration" of the system changes "very gently". Then how does this physical process help us in finding solutions to our problems? The key factor is that we will be considering problems whose optimal solutions will correspond to minimum-energy or ground states of some Hamiltonian of an Ising model. If we start with our system in the ground state (for some Hamiltonian) and we evolve it slowly enough, in other words, adiabatically, it will remain in a ground state through the whole process. But how slow is sufficiently slow? The answer is that the total time of the evolution should be inversely proportional to the square of the spectral gap, which is the minimum difference in energy between the ground state and the first excited state of the Hamiltonian.

Let us formulate the aforementioned description into a mathematical description. Suppose that H_1 is the Hamiltonian encoding the problem which we are solving; thus, its ground state encodes the optimization result we would like to find. And, suppose that H_0 is the Hamiltonian of our system. Suppose that we run the process for a total time T. The time-dependent Hamiltonian that we will consider is expressed as

$$H(t) = A(t)H_0 + B(t)H_1, \tag{6.39}$$

where A and B are real-valued functions that accept inputs over the interval $[0, T]$ such that $A(0) = B(T) = 1$ and $A(T) = B(0) = 0$. Notice that $H(0) = H_0$ and $H(T) = H_1$, exactly as we wanted. A popular choice for $A(t)$ and $B(t)$ is to set $A(t) = 1 - \frac{t}{T}$ and $B(t) = \frac{t}{T}$. Other choices that satisfy the aforementioned boundary condition are also acceptable.

As pointed out by [10], the efficiency of a given problem remains invariant whether addressed through gate-based quantum computing or adiabatic quantum computing. Consequently, the choice between these two quantum computing paradigms can be made based on the particular nature of the problem or the availability of accessing the quantum computing resources.

6.4.2 Quantum Annealing

Although we have just seen that adiabatic quantum computing is, theoretically, a perfectly viable alternative to the quantum circuit model, in its practical incarnation it is usually implemented in a restricted version called quantum annealing (QA). Originally developed for classical machines, QA is a heuristic optimization technique that applies principles from quantum mechanics to solve combinatorial optimization problems with discrete variables. It relies on the same core idea as adiabatic quantum computing, but there are some distinguishable differences. Specifically, consider the time-dependent Hamiltonian

$$H(t) = \Gamma(t)H_D + H_P, \quad t \in [0, \tau], \tag{6.40}$$

where H_P and H_D are non-commuting operators on the n-qubit Hilbert space $(\mathbb{C}^2)^{\otimes n}$ and are called problem Hamiltonian and transverse field Hamiltonian, respectively; Γ is a positive decreasing function that attenuates the contribution of H_D; and τ is the evolution time. The annealing process drives the quantum system toward the ground state of H_P, which is designed to represent the optimization problem.

QA can be physically realized by considering a network of qubits arranged on the vertices of a graph (V, E), with $|V| = n$ and whose edges E represent the couplings among the qubits. Then, the problem Hamiltonian can be defined as

$$H_P = H(\Theta) := \sum_{i \in V} \theta_i \sigma_z^{(i)} + \sum_{(i,j) \in E} \theta_{ij} \sigma_z^{(i)} \sigma_z^{(j)}, \tag{6.41}$$

where the real coefficients θ_i and θ_{ij} are arranged into the matrix Θ, and $\theta_z^{(i)}$ is the z-Pauli matrix $\sigma_z = \begin{pmatrix} 1 & 0 \\ 0 & -1 \end{pmatrix}$ on the ith tensor factor:

$$\theta_z^{(i)} = \begin{cases} \begin{pmatrix} 1 & 0 \\ 0 & -1 \end{pmatrix} & \text{on the } i\text{th vector,} \\ \begin{pmatrix} 1 & 0 \\ 0 & 1 \end{pmatrix} & \text{on other vectors.} \end{cases} \tag{6.42}$$

By definition, the set of eigenvalues of the Hamiltonian problem (6.41) is the set of all possible values of the cost function given by the energy of the well-known Ising model:

$$E(\Theta, z) = \sum_{i \in V} \theta_i z_i + \sum_{(i,j) \in E} \theta_{ij} z_i z_j, \tag{6.43}$$

where $z = (z_1, \dots, z_n) \in -1, 1^{|V|}$. In practice, the annealing procedure, also called cooling, drives the system to the ground state of $H(\Theta)$, which corresponds to the spin configuration encoding the solution:

$$z^* = \underset{z \in \{-1,1\}^{|V|}}{\arg\min}\, E(\Theta, z). \tag{6.44}$$

Given a problem, the annealer is initialized using a suitable choice of the weights Θ, and the binary variables $z_i \in -1, 1$ are physically realized by the outcomes of the measurements performed on the qubits located on the vertices V. In order to solve a general optimization problem through QA, it is first necessary to find an encoding of the objective function in terms of the cost function (6.43), which is hard in general.

However, if the quantum architecture is able to provide a fully connected graph (V, E), then any quadratic unconstrained binary optimization (QUBO) problem can be directly represented into the cost function by means of the change of variables $x_i = \frac{z_i+1}{2} \in \mathbb{B}$ where $\mathbb{B} = 0, 1$. Indeed, QUBO problems are NP-hard problems in the form of

$$\underset{x \in \mathbb{B}}{\arg\min}\, x^T Q x, \tag{6.45}$$

where Q is an upper triangular (or symmetric) matrix of real values.

The objective function in QUBO can be rewritten as

$$x^T Qx = \sum_{i=1}^{n} q_{ii}x_i^2 + \sum_{i=1}^{n}\sum_{j=i+1}^{n} q_{ij}x_ix_j$$
$$= \sum_{i=1}^{n} q_{ii}x_i + \sum_{i=1}^{n}\sum_{j=i+1}^{n} q_{ij}x_ix_j. \tag{6.46}$$

In practice, the main diagonal of Q contains the linear coefficients (q_{ii}), whereas the rest of the matrix contains the quadratic ones (q_{ij}). Although a QUBO problem is unconstrained by definition, in practice, constraints are often included in the objective function in the form of a penalty.

According to this theorem, if a quantum system's Hamiltonian changes slowly enough while the system is in its ground state, the system will remain in the ground state throughout the evolution. This concept underlies the QA process, where the system transitions from an initial simple Hamiltonian H_i to a more complex time-dependent Hamiltonian H_f, as described by Eq. (6.47):

$$H(s) = A(s)H_i + B(s)H_f, \tag{6.47}$$

where s is a normalized time factor indicating the annealing fraction, with weighting functions $A(s)$ and $B(s)$ controlled by the quantum hardware. These functions ensure that H_i gradually decreases, while H_f becomes more dominant until the system adopts a classical state when s reaches 1, leading to the measurement of the final states of the qubits.

The QUBO model employed in QA possesses the ability to represent a myriad of real-world problems, including placement [11, 12], routing [13], and scheduling [14]. It involves unipolar binary variables, meaning they can only assume values of 0 or 1, as the term **binary** indicates. Consequently, it exclusively addresses combinatorial optimization problems. The term **quadratic** pertains to the highest degree of the polynomial involved, while **unconstrained** signifies that the constraints on the variables cannot be explicitly enforced. The term **optimization** underscores that this model is employed for the purpose of minimizing or maximizing the objective function defined in (6.45)

$$H_{qubo} = \sum_{i,j,i\neq j} J_{ij}\sigma_i\sigma_j + \sum_{i} h_i\sigma_i + C, \tag{6.48}$$

where $J_{ij} = J_{ji}$ signifies the symmetric interaction coefficient or the quadratic coefficient between the ith and jth pixels, h_i represents the linear coefficient of the ith pixel, and C is a constant bias.

A QUBO problem can also be written as

$$\text{minimize/maximize} \quad y = \mathbf{x}^t \cdot Q \cdot \mathbf{x}, \tag{6.49}$$

where **x** is a vector of binary variables (for example, [0, 1, 1, 0, 1]) and Q is a matrix of constants encapsulating the problem's details. Q may either be symmetric or in an upper triangular configuration.

6.5 Quantum Inspiration for Classical Algorithms

This section presents a quantum-inspired framework to transform classical algorithms into their quantum-based counterparts. The flowchart for the quantum-inspired framework is illustrated in Figure 6.6. The framework takes the setup parameters for the population as well as the algorithm as its inputs. In addition, the stop condition is chosen as the maximum number of iterations for each optimization process. When the stop condition is satisfied, optimal values for the design variables are produced as the output.

Figure 6.6 Flowchart of the quantum-inspired framework.

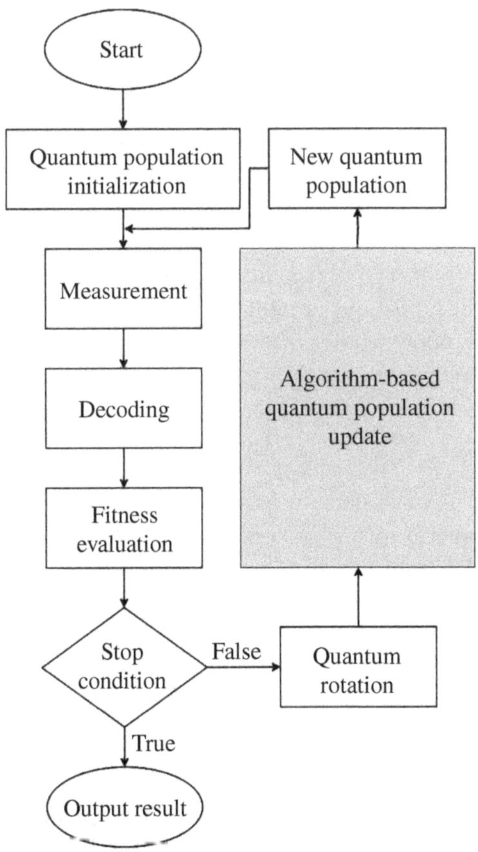

In Figure 6.6, all the steps except “Algorithm-based quantum population update” are the common steps applied to any quantum-inspired algorithm. The exception, “Algorithm-based quantum population update”, should be customized based on the classical algorithm’s characteristic features.

Each step in the flowchart in Figure 6.6 is clarified as follows.

6.5.1 Quantum Population Initialization

A quantum population consists of many quantum individuals. The ith individual is defined by an n-qubit sequence as

$$I\langle Q\rangle^i = \begin{pmatrix} \alpha_1 & \alpha_2 & \cdots & \alpha_j & \cdots & \alpha_n \\ \beta_1 & \beta_2 & \cdots & \beta_j & \cdots & \beta_n \end{pmatrix}_i = \sum_j c_i|\psi_j\rangle, \tag{6.50}$$

where $|\psi_j\rangle = \begin{pmatrix} \alpha_j \\ \beta_j \end{pmatrix}$ is the jth element of the ith individual.

Thus, a quantum population of m n-qubit individuals can be written as

$$\begin{aligned} P\langle Q\rangle &= \{I\langle Q\rangle^1, I\langle Q\rangle^2, \ldots, I\langle Q\rangle^i, \ldots, I\langle Q\rangle^m\} \\ &= \left\{ \begin{pmatrix} \alpha_{11} & \cdots & \alpha_{j1} & \cdots & \alpha_{n1} \\ \beta_{11} & \cdots & \beta_{j1} & \cdots & \beta_{n1} \end{pmatrix}_1, \begin{pmatrix} \alpha_{12} & \cdots & \alpha_{j2} & \cdots & \alpha_{n2} \\ \beta_{12} & \cdots & \beta_{j2} & \cdots & \beta_{n2} \end{pmatrix}_2, \ldots, \right. \\ &\quad \left. \begin{pmatrix} \alpha_{1i} & \cdots & \alpha_{ji} & \cdots & \alpha_{ni} \\ \beta_{1i} & \cdots & \beta_{ji} & \cdots & \beta_{ni} \end{pmatrix}_i, \ldots, \begin{pmatrix} \alpha_{1m} & \cdots & \alpha_{jm} & \cdots & \alpha_{nm} \\ \beta_{1m} & \cdots & \beta_{jm} & \cdots & \beta_{nm} \end{pmatrix}_m \right\}. \end{aligned}$$

To initialize a quantum population, first and foremost, it is required to equalize the probability amplitude of each element of the individuals. This can be achieved by applying the Hadamard gate to an array of $|0\rangle$. Each element of the individuals is expressed as

$$H|0\rangle = \frac{1}{\sqrt{2}} \begin{pmatrix} 1 & 1 \\ 1 & -1 \end{pmatrix} \begin{pmatrix} 1 \\ 0 \end{pmatrix} = \frac{1}{\sqrt{2}} \begin{pmatrix} 1 \\ 1 \end{pmatrix}. \tag{6.51}$$

The last step involves quantum rotation using the rotation matrix of Eq. (6.6). It is performed with a random angle $\theta \in (0; \pi/2]$ for each element of the individuals. When the initialization step completes, the jth element of the ith individual in the quantum population is given by

$$\begin{pmatrix} \alpha_{ji} \\ \beta_{ji} \end{pmatrix} = \begin{pmatrix} \cos(\theta_{ji}) & -\sin(\theta_{ji}) \\ \sin(\theta_{ji}) & \cos(\theta_{ji}) \end{pmatrix} \frac{1}{\sqrt{2}} \begin{pmatrix} 1 \\ 1 \end{pmatrix}. \tag{6.52}$$

6.5.2 Measurement

When the measurement is performed in the standard basis for each element, which is in the form of a qubit, of the individuals in the quantum population, these qubits will collapse to either $|0\rangle$ or $|1\rangle$, and no longer be in superposition states. In other words, quantum measurement transforms a quantum population into a classical population.

The measurement in the standard basis is expressed by the collapse of the wave function as

$$\begin{cases} b_{ji} = 0, & \text{if } p(\alpha) \leq \left|\alpha_{ji}\right|^2, \\ b_{ji} = 1, & \text{otherwise,} \end{cases} \tag{6.53}$$

where $p(\alpha)$ is a random number in the range [0; 1) and b_{ji} is the jth element of the ith individual in the classical population. To maintain the probabilistic characteristics of quantum observation, it is necessary that $p(\alpha)$ not be fixed to a constant value as in [2, 15] to adhere to quantum computing's concepts and hence ensure the quantum-inspired framework's performance as well.

Therefore, after performing measurements on the whole quantum population, the classical population is obtained with the vector representation as

$$\begin{aligned} P\langle C\rangle &= \{I\langle C\rangle^1, I\langle C\rangle^2, \ldots, I\langle C\rangle^i, \ldots, I\langle C\rangle^m\} \\ &= \Big\{ \left(b_{11} \ \ldots \ b_{j1} \ \ldots \ b_{n1}\right)_1, \left(b_{12} \ \ldots \ b_{j2} \ \ldots \ b_{n2}\right)_2, \ldots, \\ &\quad \left(b_{1i} \ \ldots \ b_{ji} \ \ldots \ b_{ni}\right)_i, \ldots, \left(b_{1m} \ \ldots \ b_{jm} \ \ldots \ b_{nm}\right)_m \Big\}. \end{aligned}$$

6.5.3 Decoding

Each binary bit sequence $\left(b_{1i} \ \ldots \ b_{ji} \ \ldots \ b_{ni}\right)_i$ corresponding to a typical individual is obtained after performing quantum measurement in the computational basis. They can then be decoded into a decimal number to represent a value of the optimization variables.

The formula for the decoding step is given by

$$value = \text{LB} + \frac{decimal}{2^n} \times (\text{UB} - \text{LB}), \tag{6.54}$$

where *value* is the real number of the optimization variable, *decimal* is the unsigned decimal value of the binary sequence $\left(b_{1i} \ \ldots \ b_{ji} \ \ldots \ b_{ni}\right)_i$, and LB and UB are the lower and upper bounds of the optimization variables.

6.5.4 Fitness Evaluation

The step of fitness evaluation aims to find the fitness values of all individuals in the population based on a predetermined fitness function (or objective function). Depending on the application or the performance metrics to be optimized, the fitness function can be selected differently. For any application, the fitness function plays a crucial role as the criterion to evaluate the "fitness" or quality of potential solutions, individuals or chromosomes in specific, within the population. As a result, the optimization process can be trained in a fitness-oriented manner to ensure better solutions at every iteration.

6.5.5 Quantum Rotation

Quantum rotation utilizes the transformation matrix in Eq. (6.6) of the rotation gate. Its purpose is to approximate the state of the currently considered individual *current* to that of the currently best individual over the completed iterations *best*.

As illustrated in Figure 6.7, this gate either increases or decreases the probability amplitude of the elements toward the individual with the best fitness.

Hence, the update of a typical quantum state, specifically the jth element of an arbitrary individual, is written as

$$\begin{pmatrix} \alpha_j^{t+1} \\ \beta_j^{t+1} \end{pmatrix} = \begin{pmatrix} \cos(\triangle\theta_j) & -\sin(\triangle\theta_j) \\ \sin(\triangle\theta_j) & \cos(\triangle\theta_j) \end{pmatrix} \begin{pmatrix} \alpha_j^{t} \\ \beta_j^{t} \end{pmatrix}, \tag{6.55}$$

where $\begin{pmatrix} \alpha_j^{t} \\ \beta_j^{t} \end{pmatrix}$ and $\begin{pmatrix} \alpha_j^{t+1} \\ \beta_j^{t+1} \end{pmatrix}$ are the quantum states before and after being updated.

Table 6.1 is a look-up table that shows the rotation scheme of each element of the current individual with respect to that of the best individual.

In fact, the $\triangle\theta$ value directly affects the evolution rate of the population. If $\triangle\theta$ is too large, the algorithm may converge to a local optimal solution; if $\triangle\theta$ is too small, the convergence rate of the algorithm will be slow [16]. In other words, $\triangle\theta$ influences the convergence speed of the algorithm. If $\triangle\theta$ is a constant value throughout the algorithm, depending on the optimization problem, a suitable value for $\triangle\theta$ is essential, which might be a time-consuming trial-and-error

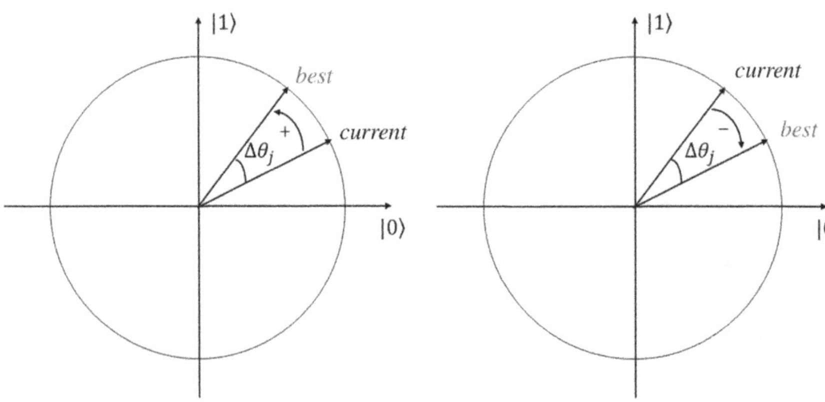

Figure 6.7 Quantum rotation toward the best individual.

Table 6.1 Lookup table for the rotation angle.

$current_j$	$best_j$	$f(current_j) \geq f(best_j)$	$\triangle\theta_j$
0	0	False	0
0	0	True	0
0	1	False	$+\triangle\theta$
0	1	True	0
1	0	False	$-\triangle\theta$
1	0	True	0
1	1	False	0
1	1	True	0

process. Therefore, instead of fixing $\triangle\theta$ as in [17, 18], an adaptive strategy for the rotation angle is applied to prevent premature convergence and improve the convergence speed. This adjusting mechanism is given by [16] as

$$\triangle\theta = \left[1 - \frac{f(current_j)}{f(best_j)}\right] \triangle\theta_0, \tag{6.56}$$

where $\triangle\theta_0$ is the coefficient of rotation angle.

6.5.6 Algorithm-Based Quantum Population Update

The steps described in Sections 6.5.1–6.5.5 are fixed for formulating any quantum-inspired algorithm from their classical versions. The only exception is the step "Algorithm-based Quantum Population Update" of Section 6.5.6—this step requires a customized update scheme for each algorithm according to its characteristic features.

For quantum-inspired algorithms, the quantum population update is a combination of quantum computing concepts and the population update process of their corresponding classical algorithms. Thus, to transform from a classical population update to its quantum counterpart, we need to take notice of (i) the qubit representation of the quantum population for both the input and output of the population update step, and (ii) appropriate adjustments to the characteristic population update equations and substeps, if necessary.

6.5.7 Quantum Genetic Algorithm

This section presents the substeps of the "Algorithm-based Quantum Population Update" step to transform the classical GA version into its quantum-inspired quantum genetic algorithm (QGA) counterpart.

6.5.7.1 Selection

The selection step plays an essential role in our QGA's implementation by determining the composition of the next generation based on the fitness of individuals in the current population. Based on the concept of natural selection, this process ensures the preferential reproduction of individuals with higher fitness, with a view to reproducing promising solutions as well as improvements over generations. In this chapter, we employ the tournament selection mechanism to take full advantage of the above-mentioned merits, which improves the algorithm in comparison with previous studies [2, 17], where the selection process was not used. This implementation is expected to contribute to the efficiency of the QGA in terms of finding optimal or near-optimal solutions.

6.5.7.2 Quantum Crossover

The quantum crossover operator involves the permutation between two randomly chosen qubits, with a specific crossover rate. Two random quantum individuals are considered before measurement, which can be expressed as

$$I\langle Q\rangle^k(t) = \begin{pmatrix} \alpha_{1k} & \cdots & \alpha_{jk} & \cdots & \alpha_{nk} \\ \beta_{1k} & \cdots & \beta_{jk} & \cdots & \beta_{nk} \end{pmatrix}_k, I\langle Q\rangle^l(t) = \begin{pmatrix} \alpha_{1l} & \cdots & \alpha_{jl} & \cdots & \alpha_{nl} \\ \beta_{1l} & \cdots & \beta_{jl} & \cdots & \beta_{nl} \end{pmatrix}_l \tag{6.57}$$

where $I\langle Q\rangle^k(t)$ and $I\langle Q\rangle^l(t)$ are the kth and lth quantum individual before quantum crossover, respectively.

If the crossover probability is satisfied, performing quantum crossover at the jth element, two new quantum individuals can be obtained as

$$I\langle Q\rangle^k(t+1) = \begin{pmatrix} \alpha_{1k} & \cdots & \alpha_{jl} & \cdots & \alpha_{nl} \\ \beta_{1k} & \cdots & \beta_{jl} & \cdots & \beta_{nl} \end{pmatrix}_k, I\langle Q\rangle^l(t+1) = \begin{pmatrix} \alpha_{1l} & \cdots & \alpha_{jk} & \cdots & \alpha_{nk} \\ \beta_{1l} & \cdots & \beta_{jk} & \cdots & \beta_{nk} \end{pmatrix}_l \tag{6.58}$$

where $I\langle Q\rangle^k(t+1)$ and $I\langle Q\rangle^l(t+1)$ are the kth and lth quantum individual after quantum crossover, respectively.

6.5.7.3 Quantum Mutation

The quantum mutation is performed using the transformation matrix of the Pauli X gate as in Eq. (6.7) with a certain mutation rate. If the mutation probability satisfies the criteria, the result of mutating the jth gene of the ith chromosome is written as

$$\begin{pmatrix} \alpha_{ji}^{t+1} \\ \beta_{ji}^{t+1} \end{pmatrix} = \begin{pmatrix} 0 & 1 \\ 1 & 0 \end{pmatrix} \begin{pmatrix} \alpha_{ji}^{t} \\ \beta_{ji}^{t} \end{pmatrix} = \begin{pmatrix} \beta_{ji}^{t} \\ \alpha_{ji}^{t} \end{pmatrix}, \tag{6.59}$$

where $\begin{pmatrix} \alpha_{ji}^{t+1} \\ \beta_{ji}^{t+1} \end{pmatrix}$ is the quantum state of the gene after being mutated.

From Eq. (6.59), it is clear that quantum mutation swaps the probability amplitudes of the mutated qubits.

6.5.8 Hybrid Quantum Firefly–Genetic Algorithm

The hybrid quantum firefly-genetic algorithm (HQFGA) is formulated mainly based on the principles of the FA integrated with quantum computing concepts. The reason for the hybrid with GA is that the HQFGA utilizes the quantum mutation operator for the random movement of the best individual in the population, which helps the algorithm avoid being trapped at local minima. The reason for the choice of the quantum mutation operator is that its high level of exploration matches the purpose of the random component for the currently best individual.

The algorithm-based quantum population update of the HQFGA can be described by Algorithm 6.2.

Algorithm 6.2: Hybrid Quantum Firefly—Genetic Algorithm (HQFGA)

```
for p = 1 : m do
    for q = 1 : m do
        if f_obj(I⟨Q⟩^p) > f_obj(I⟨Q⟩^q) then
            Update I⟨Q⟩^q based on Eqs. (6.60)–(6.62);
        Update currently best individual I⟨Q⟩* using quantum mutation;
```

In Algorithm 6.2, $f_{obj}(I\langle Q\rangle^p)$ and $f_{obj}(I\langle Q\rangle^q)$ are the respective fitness values of the pth individual $I\langle Q\rangle^p$ and the qth individual $I\langle Q\rangle^q$ of the population, and $I\langle Q\rangle^*$ is the currently best individual of the population.

Accordingly, the update of location for quantum individuals with lower attractiveness is obtained as follows [19]:

$$I\langle Q\rangle^q(t+1) = \begin{pmatrix} \frac{1}{\sqrt{\alpha_{1q}^2+\beta_{1q}^2}} & \cdots & \frac{1}{\sqrt{\alpha_{nq}^2+\beta_{nq}^2}} \\ \frac{1}{\sqrt{\alpha_{1q}^2+\beta_{1q}^2}} & \cdots & \frac{1}{\sqrt{\alpha_{nq}^2+\beta_{nq}^2}} \end{pmatrix} \odot [I\langle Q\rangle(t) + e^{-\gamma_1(\|R_1\|)^2}R_1 + e^{-\gamma_2(\|R_2\|)^2}R_2], \tag{6.60}$$

$$R_1 = I\langle Q\rangle^p(t) - I\langle Q\rangle^q(t), \tag{6.61}$$

$$R_2 = I\langle Q\rangle^*(t) - I\langle Q\rangle^q(t), \tag{6.62}$$

where $I\langle Q\rangle^q(t+1)$ and $I\langle Q\rangle^q(t)$ are the new and old locations of the attracted quantum firefly; R_1 and R_2 are the distances between the considered quantum firefly;

and γ_1 and γ_2 are the light absorption coefficients in the range [0.01, 100] that determine the convergence speed of the algorithm [19].

It is worth noticing several important terms of Eq. (6.60) for the update of quantum firefly:

- The symbol $\odot$ denotes the Hadamard product (Section 6.1.4) with the normalization factor $\sqrt{\alpha_{iq}^2 + \beta_{iq}^2}(i \in \{1, \ldots, m\})$. This is used to satisfy the condition that the sum of probabilities that a quantum state collapses to $|0\rangle$ and $|1\rangle$ equals 1.
- $(\|R_1\|)^2$ and $(\|R_2\|)^2$ are the Frobenius norm of the matrices R_1 and R_2 calculated from Eqs. (6.61) and (6.62), correspondingly.
- $e^{-\gamma_1(\|R_1\|)^2} R_1$ demonstrates the attractiveness strength between quantum fireflies p and q, while $e^{-\gamma_2(\|R_2\|)^2} R_2$ illustrates the attractiveness strength between the best quantum firefly in the current population versus quantum firefly q. From the aforementioned expressions, it can be inferred that the larger the distance that the two quantum fireflies possess, the less attraction they have for each other.

6.6 Case Study: Auto-Adjusting Hybrid Quantum Genetic Algorithm for Two-Stage Op-Amp Design[1]

6.6.1 Initialization Steps for Optimization

Regarding the two-stage Miller-compensated op-amp, its optimization process is initialized in the same way as described in Section 4.5.2 with respect to the choice of the FoM function, setup conditions, optimization variables, and design specifications.

Two algorithms were assigned to perform the sizing procedure on this opamp, namely hybrid quantum genetic algorithm (HQGA), and auto-adjusting hybrid quantum genetic algorithm (AHQGA), which implements quantum computing in several steps of the classical GA. For these hybrid algorithms, the quantum chromosome representation specifies 16 qubits, and the population size is set to 16 individuals. While crossover has a high probability, the mutation probability is typically low. As shown by [21], the rates for crossover (r_{cross}) and mutation (r_{mut}) fall within the ranges [0.8, 0.95] and [0.001, 0.05], respectively.

6.6.2 Results and Discussion

In our study, simulation results showed that the highest value for the fitness function in Eq. (4.29) was 0.5654, which was achieved with $r_{cross} = 0.8$, $r_{mut} = 0.05$, and

1 This section has been published as part of [20].

Table 6.2 Optimum design parameters achieved for the two-stage Miller-compensated op-amp with their corresponding specifications.

Design parameters	Specification	Optimum value
A_V (dB)	> 50	50.58
UGB (MHz)	> 50	61.51
PM (°)	> 60	68.75
CMRR (dB)	> 50	51.73
Power (μW)	< 250	193.78
SR (V/μs)	> 50	61.20

$\triangle\theta_0 = 0.005\pi$. With all specifications satisfied, the fitness value of 0.5654 includes the optimum design parameters, as presented in Table 6.2.

Moreover, fitness values over 100 iterations of the AHQGA are illustrated in Figure 6.8. Although the values for various op-amp design parameters may vary unpredictably throughout the iterations, their corresponding fitness values followed a monotonous upward trend, as in Figure 6.8. This conforms to the

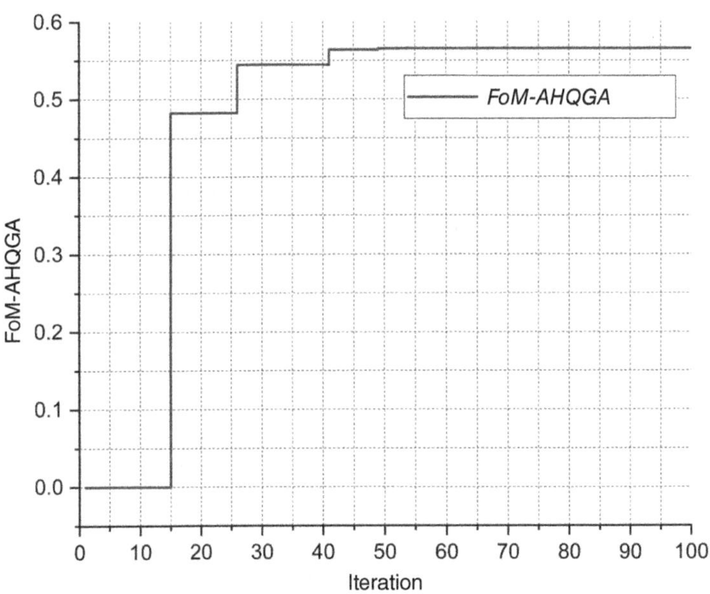

Figure 6.8 FoM values by AHQGA versus 100 iterations.

algorithm's principle of survival of the fittest which produces better individuals for each generation.

Regarding the optimization variables of AHQGA, it is worth taking into account the process grid of 5 nm of the TSMCN65 process. Therefore, the widths and lengths of MOSFETs after the optimization process needed to be rounded to reasonable values. The post-optimization sizes of all MOSFETs in the circuit are presented in Table 6.3, while Table 6.4 outlines the performance comparison between the two-stage Miller-compensated op-amp presented in this section and other studies.

Table 6.3 Post-optimization sizing results of optimized two-stage Miller-compensated op-amps.

Device	Size (W/L)
M_0, M_1	1.47 μm/0.32 μm
M_2, M_3	0.85 μm/0.15 μm
M_4	2.45 μm/0.55 μm
M_5	18.02 μm/0.55 μm
M_6	16.38 μm/0.25 μm
M_7	2.45 μm/0.55 μm
C_C	0.28 pF

Table 6.4 Performance summary of the proposed two-stage op-amp versus those of other different studies.

Performance metrics	[22]	[23]	[24]	[25]	This work
CMOS process (nm)	65	65	180	180	65
Supply voltage (V)	1.1	1.1	1.8	1.8	1.2
C_L (pF)	0.2	10	10	1	1
A_V (dB)	21.6	43	75.62	75.35	50.58
UGB (MHz)	169.7	28	16.78	45.11	61.51
PM (°)	62.4	60	62.48	61.24	68.75
CMRR (dB)	35.5	NA	NA	NA	51.73
Power (μW)	89	321	220	123.156	193.78
SR (V/μs)	288	NA	15.88	NA	61.20
FoM (HzF/A)	0.1432	0.9595	1.5214	0.6936	0.5654

From Table 6.4, in comparison to [22], it is evident that our proposed op-amp exhibits surpassing values for FoM as well as A_V, PM, and $CMRR$. Since our A_V is more than twice that of [22] (50.58 dB versus 21.6 dB), the amount of current in the circuit should be greater, primarily resulting in our higher value of power consumption. Moreover, since our work's C_C, C_L are larger (0.28 pF, 1 pF versus 0.06 pF, 0.2 pF, respectively), better *UGB* and *SR* from [22] are reasonable since they are inversely proportional to the mentioned capacitances. Regarding [23], the op-amp of our work achieves far better A_V, *UGB*, and *PM*. However, the *FoM* value is not as good, mainly due to the use of a ten-fold output capacitor C_L of [23]. This requires a larger current through the op-amp to drive the load C_L though, leading to a power consumption virtually twice as much as that of our work, which is not covered in the FoM formula.

In addition, our *UGB* as well as *PM* results dominate those of [24] and [25]; our value for *UGB* is approximately 4 and 1.5 times better, respectively. Meanwhile, the voltage gain A_V in our work is around 25 dB lower, most likely because of the higher range for the transistors' lengths as well as the higher supply voltage for the 180 nm process compared to the 65 nm process. Furthermore, the comparison of [24] with our work regarding the metrics of power consumption bears a strong resemblance to [23], with a larger C_L and corresponding higher power. Additionally, the lower power dissipation value of [25] versus that of our work might be attributed to the abundant room for the 1.8 V supply voltage that does not demand much current through the circuit. In general, although our work surpasses [24] and [25] in the *UGB* and *PM* aspects, lower FoM was observed. This results from the difference in technology node (180 and 65 nm) and lower supply voltage (1.2–1.8 V), which made it harder to design the circuit. Nevertheless, our work still proves its efficiency and feasibility based on other performance metrics.

With the purpose of further assessment of the AHQGA, Figure 6.9 illustrates comparisons of the proposed *FoM*, accompanied by that of both HQGA with fixed $\triangle\theta = 0.025\pi$ and GA. In order for the analysis to be judicious, all optimization algorithms were set up with the same initial conditions, including a similar initial population, crossover, and mutation rate ($r_{cross} = 0.8$ and $r_{mut} = 0.05$) as well as the number of iterations (100 in total). Table 6.5 provides further deliberations on the optimization results of the three mentioned algorithms.

Table 6.5 Optimal solutions from proposed AHQGA, HQGA with fixed $\triangle\theta$, and traditional GA.

Criterion	AHQGA	HQGA	GA
Optimal FoM value	0.5654	0.5581	0.5916
Iterations to reach optimal FoM value	49	44	94

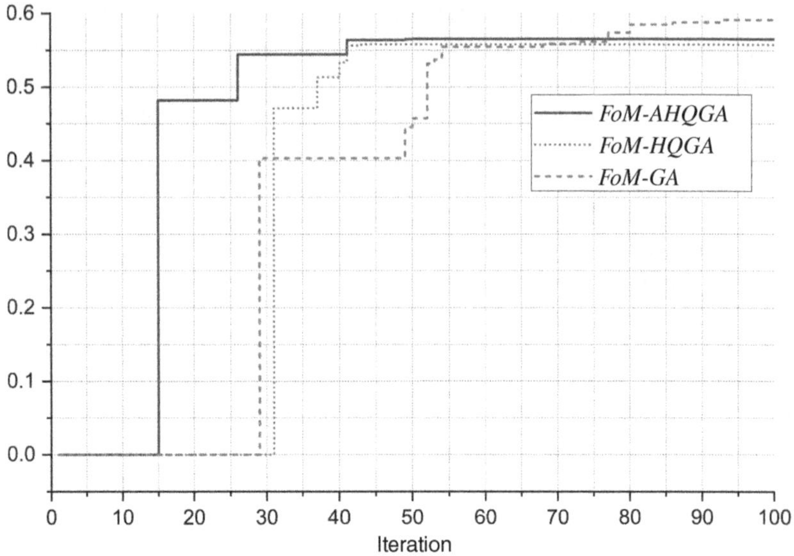

Figure 6.9 FoM comparison between AHQGA, HQGA with fixed $\triangle\theta$, and GA over 100 iterations.

According to Table 6.5 and Figure 6.9, AHQGA dominates its HQGA counterpart for the optimal value of FoM, where AHQGA achieves the 0.5654 value compared to 0.5518 of HQGA. Moreover, escaping from the so-called "FoM-zero" search space, which occurs in solutions that meet saturation conditions but do not satisfy specifications as in (4.28), should be considered essential as this will prevent the optimization algorithm from being trapped in this undesirable search space. AHQGA was successful in this respect at the 15th iteration, which was two times faster than HQGA at around the 30th iteration. In other words, AHQGA completely prevails over HQGA, which corresponds to the purpose of the auto-adjusting scheme as an upgrade for HQGA.

Lastly, it is undeniable that AHQGA is inferior to GA with respect to the optimal FoM value (0.5654 versus 0.5916). However, two key points must be considered in this comparison. First, similar to HQGA, AHQGA exceeded GA concerning the required iterations to get rid of the "FoM-zero" search space by the same two-fold ratio. Second, the number of iterations to reach the optimal solution of AHQGA far outnumbered that of GA, namely 49 versus 94 iterations. In summary, the two above-mentioned benefits of AHQGA over GA might compensate for the slightly lower optimal FoM value.

References

1 M. A. Nielsen and I. L. Chuang, *Quantum Computation and Quantum Information: 10th Anniversary Edition.* Cambridge: Cambridge University Press, 2010.
2 R. Lahoz-Beltra, "Quantum Genetic Algorithms for Computer Scientists," *Computers*, vol. 5, p. 24, 2016.
3 H. Y. Wong, *Introduction to Quantum Computing*. Cambridge: Springer, 2010.
4 R. A. Horn and C. R. Johnson, *The Hadamard Product*. Cambridge University Press, 1991, pp. 298–381.
5 F. Rothlauf, *Optimization Methods*. Berlin, Heidelberg: Springer Berlin Heidelberg, 2011, pp. 45–102.
6 IBM-Quantum-Learning, "Reference states," accessed 2025-01-23. [Online]. Available: https://learning.quantum.ibm.com/course/variational-algorithm-design.
7 S. Pattanayak, *Quantum Machine Learning with Python Using Cirq from Google Research and IBM Qiskit*. Cambridge: Springer, 2021.
8 D. Dong, C. Chen, H. Li, and T.-J. Tarn, "Quantum Genetic Algorithms for Computer Scientists," *IEEE Trans. Syst. Man Cybern. Part B Cybern.*, vol. 38, pp. 1207–1220, 2008.
9 P.-H. Fang, Y.-S. Chen, J.-S. Wu, and P. Yu, "Inverse Reticle Optimization With Quantum Annealing and Hybrid Solvers," *IEEE Access*, vol. 12, pp. 33069–33078, 2024.
10 A. D. Meglio, E. F. Combarro, S. González-Castillo, and A. D. Meglio, *A Practical Guide to Quantum Machine Learning and Quantum Optimization: Hands-on Approach to Modern Quantum Algorithms*. Packt Publishing, 2023.
11 B. Dury and O. Di Matteo, "A QUBO formulation for qubit allocation," *arXiv preprint arXiv:2009.00140*, 2020.
12 S. Speziali, F. Bianchi, A. Marini, L. Menculini, M. Proietti, L. F. Termite, A. Garinei, M. Marconi, and A. Delogu, "Solving sensor placement problems in real water distribution networks using adiabatic quantum computation," in *Proc. IEEE Int. Conf. Quantum Comput. and Eng. (QCE)*, Oct. 2021, pp. 463–464.
13 T. D. Tambunan, A. B. Suksmono, I. J. M. Edward, and R. Mulyawan, "Quantum annealing for vehicle routing problem with weighted segment," in *AIP Conference Proceedings*, vol. 2906. AIP Publishing, 2023.
14 K. Ikeda, Y. Nakamura, and T. S. Humble, "Application of Quantum Annealing to Nurse Scheduling Problem," *Sci. Rep.*, vol. 9, no. 1, p. 12837, Sep. 2019.

15 Y. Sun and M. Ding, “Quantum genetic algorithm for mobile robot path planning,” in *2010 Fourth International Conference on Genetic and Evolutionary Computing*, 2010, pp. 206–209.

16 Z. Liu, X. Li, J. Jiang, and S. Wang, “A novel improved quantum genetic algorithm for robot coalition problem,” in *2016 IEEE International Conference on Information and Automation (ICIA)*, 2016, pp. 2061–2064.

17 D. Li, J. Zhao, H. Zhang, P. Qiao, and J. Zhuang, “An improved many worlds quantum genetic algorithm,” in *2015 11th International Conference on Natural Computation (ICNC)*, 2015, pp. 210–214.

18 K.-H. Han and J.-H. Kim, “Quantum-Inspired Evolutionary Algorithm for a Class of Combinatorial Optimization,” *IEEE Trans. Evol. Comput.*, vol. 6, no. 6, pp. 580–593, 2002.

19 F. Zitouni, S. Harous, and R. Maamri, “A Novel Quantum Firefly Algorithm for Global Optimization,” *Arabian J. Sci. Eng.*, vol. 46, pp. 86–100, Apr. 2021.

20 T. Q. Do, H. T. Nguyen, B. D. E. McNiven, T. Hoang, L. Zhang, D. A. Octavia, and T. Q. Duong, “An Auto-Adjusting Hybrid Quantum Genetic Algorithm-Spectre Platform for the Multi-Objective Optimization of Analog Circuit Sizing,” *Alexandria Eng. J.*, vol. 116, pp. 574–586, Mar. 2025.

21 X.-S. Yang, S. F. Chien, and T. O. Ting, “Chapter 1 - bio-inspired computation and optimization: An overview,” in *Bio-Inspired Computation in Telecommunications*, Boston, MA, England: Morgan Kaufmann, 2015, pp. 1–21.

22 R. Rashid and N. Nambath, “Area optimisation of two stage Miller compensated op-amp in 65 nm using hybrid PSO,” *IEEE Trans. Circuits Syst. II Exp. Briefs*, vol. 69, no. 1, pp. 199–203, Jan. 2022.

23 M. El-Sisi and M. Dessouky, “Procedural analog design automation using building block optimization,” in *Proc. IEEE International Conference on Electronics, Circuits, and Systems (ICECS)*, Cairo, Egypt, Dec. 2015, pp. 474–477.

24 P. Das and B. Jajodia, “Design automation of two-stage operational amplifier using multi-objective genetic algorithm and SPICE framework,” in *Proc. Int. Conf. Invent. Comput. Technol. (ICICT)*, Nepal, Jul. 2022, pp. 166–170.

25 S. A. Abdelaal, A. Hussein, and H. Mostafa, “A Bayesian optimization framework for analog circuits optimization,” in *Proc. 15th Int. Conf. Comput. Eng. Syst. (ICCES)*, Cairo, Egypt, Dec. 2020, pp. 1–4.

Index

Advanced Techniques for Optimal Sizing of Analog Integrated Circuits: Quantum Computing, Machine Learning, and Bio-inspired Optimization, First Edition.
Trang Hoang, Thinh Quang Do, Thang Quoc Nguyen, Hoang Trong Nguyen, Lihong Zhang, Octavia A. Dobre, and Trung Q. Duong.